Struggles for Climate Justice

Brandon Barclay Derman

Struggles for Climate Justice

Uneven Geographies and the Politics of Connection

Brandon Barclay Derman
Department of Environmental Studies
University of Illinois at Springfield
Springfield, IL, USA

ISBN 978-3-030-27964-6   ISBN 978-3-030-27965-3   (eBook)
https://doi.org/10.1007/978-3-030-27965-3

This Palgrave Macmillan imprint is published by the registered company Springer Nature Switzerland AG.
The registered company address is: Gewerbestrasse 11, 6330 Cham, Switzerland

In memory of my grandparents

Acknowledgments

I owe the activists, advocates, and governmental staff who generously made time to answer my interview questions for the inspiration that has guided this book and much of what insight it contains. Its deficits, of course, are mine. Many of those whose voices and visions inform the text most meaningfully must remain anonymous. Broader thanks are also due to them, however. The individuals I interviewed and whose work I observed in making this book devote their working lives and often much more to the cause of a more just and sustainable future for us all.

Before this project began, mentors and friends at Hunter College of the City University of New York opened the doors that led onward to it. Among them are Jochen Albrecht, Laxmi Ramasubramanian, and Haydee Salmun. The work would still not have begun or gathered necessary steam, however, without many conversations at the University of Washington (UW) among those connected with Three Degrees Warmer and its offshoots: Jeni Krencicki Barcelos, Jenn Marlow, David Battisti, Stephen Gardiner, Greg Hicks, Josh Griffin, and students in multiple years of the Climate Justice Seminar.

My doctoral adviser, Steve Herbert, and committee members Vicky Lawson, Matt Sparke, and Michael McCann, as well as Sarah Elwood, Lucy Jarosz, and other supporters in Geography and Law, Society and Justice at UW saw promise in a geographic/socio-legal study of climate justice, and provided steadfast support and guidance on what felt like a long road. They also helped build bridges by which that study gained readers and allies, and without which this text would have been less grounded and I less committed. To Asuncion St. Claire, Siri Gloppen, Kjersi Fløttum, Patrick Bond, Jackie Dugard, Susan Sterett, Anna-Maria Marshall, and Stephen Gardiner, who all convened workshops and/or edited related collections, and the many others who participated in Bergen, Durban, Oñati, Boulder, and Seattle: thank you.

Among the fruit of those meetings, the editors of the *Oñati Socio-legal Series* graciously allowed me to borrow from my paper "Revisiting Limits to Legal Mobilization for Global Climate Justice" in what became Chap. 2. Similarly, Taylor & Francis granted generous permission to draw from my chapter in the *Routledge Handbook of Climate Justice*, edited by Tahseen Jafry with Michael Mikulewitz and Karen Helwig, in parts of Chap. 5.

I could not have visited the places or heard from the people who articulated the challenges and visions of climate justice chronicled in these pages were it not for funding from the United States National Science Foundation (Grant number 1129127), the European Union Centers of Excellence, the UW Department of Geography and President's office, and the College of Public Affairs and Administration and Provost's office at the University of Illinois Springfield (UIS). The voyages they facilitated would have been hard ones without the intellectual and personal comradery of Sonja Klinsky, Shangrila Joshi Wynn, Joel Wainwright, Lisa Vanhala, and Tracey Osborne; the work of those who make *Democracy Now*'s coverage of climate negotiations and alternative summits; and the warmth of the Rasch's Copenhagen home. Between data-gathering trips, my students at UW and UIS were frequent interlocutors and sounding boards for the ideas that shaped this book. Their questions and observations helped to clarify and motivate the analysis that follows, and the research assistance a few of them provided was invaluable.

At Palgrave Macmillan, Rachael Ballard initiated and sustained the conversation that resulted in this book, and Joanna O'Neill has been instrumental in its realization. I learned from and was inspired onward both by anonymous reviewers and by the careful and timely comments of Tiffany Grobelski, Katie Gillespie, Riaz Tejani, Eugene McCarthy, and Mike Babb.

My fellow travelers through the UW geography program and now at UIS make the lumpy life of research and teaching more and more enjoyable. At and after UW, Katie, Tiffany, Mike, Jesse McClelland, in particular, gave friendship as well as intellectual attention to this project in great measure. In Springfield, Richard Gillman-Opalsky, Riaz, Eugene, Megan Styles, and Bob Smith read chapter drafts and/or supplied timely advice and support. All remaining errors and oversights are, again, mine.

Closer to home, my parents and siblings, the Brewster family, and above all Anne and Lucy sustained me and improved this book with their seemingly infinite patience, levity, good sense, and love.

Contents

Abbreviations

BASIC country group	Brazil, South Africa, India, and China
CBDR/CBDR-RC/CBDR-RC-NC	Common But Differentiated Responsibilities and Respective Capabilities (in the light of different National Circumstances)
CIA/New CIA	New Cowboy Indian Alliance
CJA	Climate Justice Action
CJN!	Climate Justice Now!
COP	Conference of the Parties to the United Nations Framework Convention on Climate Change
DAPL	Dakota Access Pipeline
DLF	Democratic Left Front
ECJP	Environmental and Climate Justice Program (of the National Association for the Advancement of Colored People)
EJOLT	Environmental Justice Organizations, Liabilities and Trade
EU	European Union
FOEI/FOE	Friends of the Earth International/Friends of the Earth
G77	Group of 77 (coalition of 134 developing countries)
GDRs	Greenhouse Development Rights framework
GHG	Greenhouse Gas
GMO	Genetically Modified Organism

HRC	United Nations Human Rights Council
IACHR	Inter-American Commission on Human Rights
ICC	Inuit Circumpolar Council
ICCPR	International Covenant on Civil and Political Rights
ICESCR	International Covenant on Economic, Social and Cultural Rights
IEN	Indigenous Environmental Network
IPCC	Intergovernmental Panel on Climate Change
KP	Kyoto Protocol
KXL	Keystone XL pipeline
LDCs	Least Developed Countries
LVEJO	Little Village Environmental Justice Organization
NAACP	National Association for the Advancement of Colored People
NDCs	Nationally Determined Contributions
NGO	Non-governmental Organization
OAS	Organization of American States
OHCHR	United Nations Office of the High Commissioner of Human Rights
PACJA	Pan-African Climate Justice Alliance
REDD/REDD+	Reducing Emissions from Deforestation and forest Degradation
RWA	Rural Women's Alliance
SLR	Sea Level Rise
UKZN	University of KwaZulu-Natal
UNEP	United Nations Environment Program
UNFCCC	United Nations Framework Convention on Climate Change
WHO	World Health Organization
WPCCC	World People's Conference on Climate Change and the Rights of Mother Earth
WTO	World Trade Organization
XR	Extinction Rebellion

List of Figures

Introduction

Chattering voices drop to silence as two images appear on a screen hung over a deep stage to one side of a darkened gymnasium. On the left side of the screen, a sleeping woman stirs, taps a button on a clock, throws back a coverlet, and gets up. The room around her is cramped and modern. In the right-hand portion of the screen, a second woman rises from a low platform in semi-darkness, bare feet meeting a dirt floor. Both wordlessly begin their day. Soon the woman on the left is obscured in a shower stall, as steam rises. Meanwhile, the woman on the right emerges from a darkened hut to slanting light. Laden with pots, she joins a rutted track. At left, the woman finishes her toilette, closes a tap, and moves into a small kitchen, igniting the element beneath a kettle. The woman to the right walks on, as her path cuts through tall grass. She at left finishes her morning meal, rinses dishes, fixes hair and coat. She to the right places each foot carefully, as the path descends unevenly to the bank of a river. To the left now, small crowded rooms stand empty. To the right a woman's arms lower a large vessel into flowing water. The audience remains quiet as the images fade.

***

Later, members of the transnational peasant and farmer alliance La Via Campesina fill the stage. In the hot darkness, candles sway in time with their voices. Brown faces, embroidery, and work clothes underscore the distance many have come, from homes and fields.

***

Now the gymnasium—the largest space in the community center hosting "Klimaforum" open gatherings during United Nations (UN) climate talks in Copenhagen—thunders. A full house pounds the bleachers to the refrain of Friends of the Earth International and Nigeria leader Nnimmo Bassey's poem "We thought it was oil, but it was blood."

***

On "the inside" too—behind the glass exterior of the Bella Center complex, where world leaders and besuited staff debate binding limits on greenhouse gas emissions—the troubled human face of climate change is on display. Rumors of secret texts circulate, as thousands of young volunteers pace or recline in the corridor. The small island state of Tuvalu has led a walkout from negotiations, to the consternation of the Secretariat, and cheers from these youths. No progress will be made on an agreement today.

Accredited by an observer organization, like the volunteers, I can enter and move between both spaces, though the journey is time-consuming and unnerving. I slide past shouting protestors and armored guards at the conference center gate, toward whatever can be witnessed on the other side.

On the dais of a massive plenary hall, the chairperson reads items from a proposed agenda. Delegates deliver lengthy preliminaries, make impassioned pleas, shame one another, or keep quiet. As negotiations stall again and again, contact groups, bilaterals, and informal consultations proliferate. Most are closed to observers.

***

Saturday December 12, 2009, Global Day of Action
After lunch, my Danish hosts and I join marchers in the streets of the Old City. Dazed by the spectacle enfolding us, sternums pulsing to the bump of soundsystem trucks, we move like particles in a river, scenes coalescing and dissolving around us: "Climate Justice Now!" on placards in many lan-

guages; bicycle riding polar bears and pandas; "change the system, not the climate" in Danish marker on a bedsheet; black facemasks, boots, and hoodies; the flag of the United Nations; the image of a blue planet.

The Scandinavian night beats us to the end of the march route, across a wide field from the Bella Center. Paraffin drips from torches as Mary Robinson, former Irish President and UN human rights Commissioner, booms from a platform:

> Look at you! .... I'm told a hundred thousand of you.... We are here because we are reminding the decision makers.... [T]he world that has benefitted from a carbon development [is] ... causing great suffering: 300,000 people *died* last year from the impacts of climate change.
>
> So we want a *fair*, *ambitious*, and *binding* agreement here in Copenhagen – and binding means *legally* binding.

Moments later, Tom Goldtooth, executive director of the Indigenous Environmental Network, intones:

> [W]e came here as Indigenous peoples from every region of the earth, demanding that our rights be recognized.... *So stand up* – not only for the rights of Indigenous peoples, but *your* rights: people of the world; people of civil society....
>
> [A] lot of our world leaders are not negotiating for you, but they are negotiating for the corporations.... We are talking about systemic change.... We want real solutions, not false solutions. Nuclear power is not clean energy; GMOs and tree plantations [are] not a solution, geoengineering is not a solution, and the carbon market is not a real solution ... we have to cut the valve of a fossil fuel economy!

New speakers take the stage and cheers give way to drums, as marchers dance around torch bonfires. Ducking under a fence, I locate my conference badge and prepare to cross again between what seem separate but simultaneous worlds.

***

Nearly 2000 marchers are arrested over the next four days. A contingent of protestors clashes with police outside the Bella Center. Hundreds are detained, some physically assaulted. Later, the Danish High Court will rule that police actions during the meetings violated the European Convention of Human Rights and Fundamental Freedoms.[1]

***

Following the surge of passion and protest in Copenhagen—both within negotiations and outside—the widely anticipated fifteenth Conference of the Parties (COP15) of the United Nations Framework Convention on Climate Change (UNFCCC) concluded, on December 19, 2009, to broad condemnation.[2] Between continual complaints about secret texts and back-room deals, and the well-documented exclusion of most delegations from the closed-door session out of which the briefly sketched Copenhagen Accord finally emerged, leaders could hardly be said to have agreed on the shape of a new global climate regime. No binding cuts to greenhouse gas emissions were offered, to supplant those owed by a few industrialized countries under the Kyoto Protocol, then quickly approaching expiration. As many predicted, pledges to mitigate emissions entered later under the Accord fell far short of scientifically based targets. Further, as the two-week conference wore on, the presence and role of observers was increasingly restricted, including those of some organizations long accredited and widely recognized as contributing constructively in prior negotiations.[3] For some, these developments signaled a worrying slide into less accountable forms of governance at a crucial moment in the evolution of the international treaty, the global economy, and the climate itself.[4]

In retrospect, the disruptions of COP15 marked an apotheosis within a period of acute uncertainty for the climate convention, from which it would emerge revivified, though fundamentally altered, with the Paris Agreement of 2015. By June 2019, 185 parties to the Convention had ratified the Agreement, and all 197 had either signed, ratified, or acceded to it, implicitly endorsing its more detailed and

---

[1] McKie and Zee 2009; *The Guardian* 2009; Zee 2010; Danish Institute for Human Rights 2012; Featherstone 2012; Johnson 2012.

[2] Dire evaluations of Copenhagen's COP were voiced by many participating state parties, civil society organizations, members of the secretariat, and academic analysts. See, for example, BBC News 2009; Vidal et al. 2009. Not all evaluations of COP15 were negative, however. See, for example, Bodansky 2010, arguing that the Copenhagen outcome signaled a "potentially significant breakthrough (239)."

[3] Bas Arts 1998; Newell 2006; Betsill and Correll 2008.

[4] See Fisher 2010; Bernauer 2013; Shift Magazine 2010.

legalistic realization of the "pledge and review" system proposed in Copenhagen. In this system, every country would commit to "nationally determined contributions" to reduce greenhouse gases, and many to other means of addressing ongoing and imminent climate change impacts within their borders. What the Agreement termed "ratcheting mechanisms," consisting largely of transparency requirements and structured dialogues, would, together with negotiated transfers of finance, expertise, and technology, take the place of mandated reductions based in science and account for the vast differences in historical responsibility, vulnerability, and capacity between countries, both of which the Kyoto Protocol had sought to address in "legally binding" terms. As the second decade of the twenty-first century drew to a close—the third for international climate negotiations—aggregate commitments under the Agreement promised only meager diversion from "business as usual" emissions scenarios, which imply the need instead for a thoroughgoing transformation of the world's economies within the coming decade in order to stave off anthropogenic changes otherwise catastrophic, though decidedly uneven, for human communities as well as ecosystems: merely a waypoint, perhaps, on the pathway to an uninhabitable planet.[5]

From the outset, the general direction of international climate governance, if not its provocative revelation at COP15, could, in effect, be read off from the combination of longstanding conditions and ill-timed conjuncture. To wit, the weakness of international legal compulsion and the dominance of domestic political and economic forces, particularly those of geopolitically powerful countries, amidst the context of a global financial crisis felt widely by 2009. The latter merely added, however, to growing fears of economic harm from policies designed to address climate change, and thereby to a broader re-framing of negotiations in financial terms, which cast institutional structures and actors inherited from prior environmental treaties as increasingly inappropriate, or even irrelevant.[6] From a more structurally oriented perspective, the demise of global climate governance ambitions could be traced to the imperative of capital accumulation within a fossil-fueled energy paradigm, ably facilitated by the corporate capture of state institutions and an international system all too amenable to the realpolitik of nationally defined economic interests.[7]

---

[5] Wallace-Wells 2019.

[6] See, for example, Broder 2011; a Brussels-based NGO worker spoke of this transition in an interview in 2011; see Derman 2014.

[7] Peet et al. 2010; Bond 2011; Paterson 2011.

Despite the apparent overdetermination of Copenhagen's conflicts and the subsequent internationally sanctioned deepening of climate-related crises, however, it was also clear to many observers that a potentially global social movement had coalesced with new breadth and exposure during COP15. Gatherings in Klimaforum and other open spaces across the city, street demonstrations, and contention in the negotiating halls were noted widely for their fervor as well as their social and geographic diversity.[8] The different settings within which I witnessed those events—the securitized "inside" of negotiations and the often-chaotic "outside" of peoples' forums and actions—as well as the channels that linked them, were each integral to the rising salience of "climate justice." Ten years on, climate justice is a prominent plank in many segments of a broader burgeoning "climate movement," proclaimed by a wide range of groups as part of advocacy and activism centered on many forums, communities, and administrative scales. It is arguably much more prominent now in civil society activities within and outside UNFCCC meetings, as well as those connected with other decision-making venues. Numerous legal cases—emerging, evolving, or decided—and even a few state policies have been framed in its terms or closely related ones.

Journalistic and scholarly accounts have also emerged, documenting or analyzing elements of the "climate justice movement," or conditions and drivers of climate *in*justice.[9] Rarely, however, have such studies examined the varied landscapes of climate justice mobilization and thought, in evidence already on the Day of Action and in the negotiating halls of COP15, particularly in light of the commonalities and crosscurrents by which those efforts might somehow be, on the one hand, synergistically allied and, on the other, decisive for the diverse, far-flung settings and constituencies they target. Nor have studies typically examined this panoply of advocacy and activism in conjoined empirical and conceptual terms. That is, many entries in the growing literature on climate justice reveal or inspire, either through careful analysis of the institutional, legal, theoretical, or ethical bases for projects of climate justice or by way of forthright grounding in the trenches, networks, and principles of governance or social struggle. While building upon the insights such studies offer, this book focuses instead on the construction of "climate justice" as a stake in political, legal, and policy-oriented initiatives within and across contexts. Climate justice initiatives must, after all, be geo-historically (and often institution-

---

[8] Fisher 2010; Chatterton et al. 2013.

[9] The list grows, but touchstones include Roberts and Parks 2006; Tokar 2010; Bond 2011; Shearer 2011; Featherstone 2012; Chatterton et al. 2013; Stephenson 2015; Gardiner and Weisbach 2016; Heyward and Roser 2016; and Klinsky and Brankovic 2018.

ally) situated as well as globally synoptic, if they are to effectively contend with any of the range of specific social conditions tied to the broader temporal and spatial extents of global environmental change, which are themselves novel in relation to most political and institutional commitments.[10] In pursuing these and other themes, this study bears the marks of my own necessarily finite intellectual toolkit, assembled primarily from geography and socio-legal studies and, to a lesser extent, social movement studies. It also reflects the combination common in those fields of qualitative fieldwork, close reading, and conceptual analysis, the latter inflected by disciplinary predilections for critical and communitarian strands of social theory. Accordingly, this study is guided by questions of a particular type. What, for instance, do different climate justice initiatives share, and what differentiates them? What, in fact, does "climate justice" mean in these contexts? Moreover, what do the framing of these efforts and their progression within the settings in which they have unfolded suggest about the conditions that produce and sustain climate *in*justice, how those conditions might be unmade, and what might take their place? For the preceding decade alone witnessed a thoroughgoing, largely intentional reconfiguration of international policy as well as the rise of a global (albeit contextually differentiated) insurgent movement, even as the planet descended into potentially calamitous climate-related crises.

A cluster of unsettling encounters with climate scientists, legal scholars, and social theorists pushed me to consider such political problems raised by anthropogenic climate change and its uneven geographies, just as the reconfiguration of global climate governance appeared on the horizon. COP15 and the tumultuous energies surrounding it then provided an experiential introduction to international negotiations as well as the wider politics of climate justice. From Copenhagen I charted a path gathering material through which to understand those politics, guided methodologically by a combination of purposive and snowball sampling.[11] That path has connected locales in and around ten cities on four continents, traversing a series of international negotiations, civil society forums, and street demonstrations; the home offices of non-governmental organizations (NGOs) and governmental agencies; and various schools, tents, and makeshift kitchens where activists met and strategized. During and between those encounters, I corresponded peri-

---

[10] There is a clear connection in this political challenge with the discussion of "militant particularisms" and other terms that index the need to maintain or construct situated meaning in conjunction with wider, even global, coherence and solidarity. On this theme, Harvey 1996; Featherstone 2008; and Routledge 2011 are each formative antecedents to my interest in this aspect of climate justice politics.

[11] Babbie 2007.

odically with activists and analysts elsewhere by voice, video, and social media, and traced their work and its antecedents through many briefs, decisions, white papers, news accounts, blog posts, and other artifacts.

As I conducted interviews, digested texts, and observed interactions, the disproportionate damages of climate change assumed ever more specific and unsettling detail. Scanning across those materials and encounters, though, I also saw a set of common themes with analytical purchase. Those themes linked the drivers of climate change and climate injustice with many distinct response efforts, whether framed in terms of justice or not. That those drivers and responses ought to cut in generally opposing directions, and that responses themselves proceed along differing lines, suggests that the commonalities between them are somewhat abstract. They are domains, in effect, encompassing specific sets of relations and processes, in which different conceptions, practices, and ordering principles might apply. The drivers of climate change and related injustices can be seen as particular configurations in these domains, responses as attempts at reconfiguration. Reflecting the disciplinary predilections already mentioned, three such consequential and distinguishable - albeit interconnected - domains could be named "the socio-ecological," "the socio-spatial," and "the governmental."

Tracking the construction and fate of climate justice efforts in terms of the relations and processes they confront or further within these domains makes up much of the analysis in ensuing chapters. This relationally oriented perspective is itself developed at more length in Chap. 1. The domains and the import of configurations within them must at least be sketched, however, in the remainder of this Introduction.

The anthropogenic origins and global reach of climate change impacts illustrate in stark fashion that the conditions of human life in any one place are consequentially connected to the actions of people elsewhere, as well as the non-human entities and systems with which we interact and on which we depend. Though subject to broad and deep scientific consensus, widespread recognition across societies of these socio-spatial and socio-ecological connections is still recent in historical terms. It largely post-dates, for instance, many fundamental legal and political institutions, as well as the forms of knowledge and practice that continue to structure them. Unsurprisingly, then, many institutions and the modes of life they support do not reflect or respond coherently to the recognition and evolution of such connections—although they often reflect forms of social power which themselves depend on relations of control over human populations, space, and the non-human. These forms of power and the knowledge and practices that have long supported them are unfortunately, therefore, woven into many coordinated responses to climate change, as epistemic predilections, structuring principles, and operating procedures. Moreover, if socio-ecological ties are often elided and socio-spatial ones obscured or constrained in these responses, the forms and outcomes of such efforts are also shaped by governmental relationalities. This third domain encompasses processes, relations, and practices associated with decision making, political representation, and participation.

Because configurations of these three domains affect both perception and action, they can make the strong ties of climate change and climatic injustice both less visible and more difficult to address through avenues like legal argument, electoral politics, and policy formation at a variety of scales. It should be clear, however, that explaining climate injustice in terms of institutionalized and acculturated relationalities in these domains does not deny consequential roles for political economic or nationalist forces, among others. It opens opportunities, instead, for empirically rich accounts of how such imperatives operate in specific contexts, through particular ways of understanding and responding to climate change, with corresponding outcomes.

Clearly, in this light, climate justice thought and mobilization must seek to identify, expose, and dismantle the relationalities that have produced and perpetuated climate injustice, and proffer in their place more egalitarian, sustainable alternatives. Indeed, one of the central claims of this book is that, while strikingly diverse in its guises, climate justice is fundamentally a "politics of connection," oriented around the three domains I have just briefly described, with all of the challenges and opportunities they imply. Advocates themselves sometimes speak of their work in similar, if, typically, less abstract, terms. As one quipped, for instance, at an open venue presentation during COP24 in Katowice, Poland, in 2018: "Social policy has to be at the heart of the climate policy; it's a matter of connection."

Orientating toward relationality in slightly more abstract terms, however, supports an analysis that cuts—or rather synthesizes—across the social and geographic settings of the political projects and empirical materials this study examines. It also enables critical insight, by highlighting the constitutive connections and separations on which different responses to climate change are based, those they recognize explicitly or elide, and those they construct or erode. Finally, examining the (re)production of climate injustice and activists' responses with the level and type of abstraction provided by this multi-dimensional relational orientation enables drawing connections between an analysis of climate justice and broader discussions in scholarship and politics, with which the final chapter is concerned.

Chapter 1 develops and begins to deploy this analytical perspective, first by linking the three dimensions sketched briefly above with global and geographically disaggregated understandings of the origins and impacts of anthropogenic climate change. Together, these help to expose the relational production of what can be called climate injustice. The first chapter continues by characterizing crosscutting conditions within which advocates and activists have developed relationally attuned responses to climate injustice, providing the context for the emergence of those efforts as well as clarifying their stakes. The four subsequent chapters each examine the ways in which the relationalities involved in producing and contesting climate injustice have played consequential roles in specific settings and mobilization efforts.

The two chapters of Part I focus on efforts structured by international rules, principles, and procedures. Chapter 2 analyzes two early legal claims, by the Inuit Circumpolar Council (ICC) and government of the Maldives, which argued that anthropogenic climate change violates human rights. In different ways, both claims illustrate how legal analysis incorporating non-legal forms of knowledge can substantiate the attribution of responsibility for climate harm, which necessarily involves mechanisms of socio-ecological and socio-spatial connection. On the other hand, foundational legal categories also supported competing interpretations, which render that attribution dubious or impossible. The disjunctures between these differing assessments of responsibility illustrate the importance of knowledge categories in causal analyses, as well as the ties between socially constructed spaces, understandings of nature, and the power embodied in law. Chapter 3 shifts to attend more closely to the period of reconfiguration within the UN climate convention briefly summarized above. Between the COPs of Copenhagen and Paris, parties struggled to define key terms within the convention that would bear consequentially on the distribution of responsibility among themselves, as well as their duties to each other. That legalistically framed debate, however, was always situated within and came ultimately to express stark differences in geopolitical power between formally equal negotiating parties. The evolution of the climate convention helps to illustrate how those differences are, in effect, mutually constitutive with the institutionalized socio-spatial and governmental separations of the international system. At the same time, debates that unfolded in the course of that evolution illustrate the necessity of language in political struggle and some of the reasons why law so often supplies such a language.

Part II turns from examples of efforts for climate justice within international institutions to some of those proceeding on the transnational "outside," and in more socially and geographically specific settings. Chapter 4 examines the construction of a transnational politics of climate justice through assemblies like Klimaforum and the street actions during COP15, which were similar to others that have taken place outside other international meetings. Organizers have intentionally structured political activity in these settings through connective forms of spatial and social practice, challenging the governmental closures of official meetings and facilitating the coming together of a diverse set of affected groups and allies. The analytical work, too, by actors involved in these settings is crucially connective and again illustrates key roles of alternative forms of knowledge, in this case both for recognizing climate injustice and for putting forth meaningful socio-ecologically relational alternatives. At the same time, this "outside" work is difficult in its own ways. The chapter therefore also explores a series of material and political challenges associated with these transnationally connective practices and analyses. Chapter 5 shifts empirical focus once again, to efforts locating the objects and subjects of socio-ecological justice more precisely in social and physical space. In two parts, the chapter analyzes examples centered in the United States, considering first the work of the National Association for the Advancement of Colored People (NAACP) and its Environmental and Climate Justice Program (ECJP), and second, that of the New Cowboy Indian Alliance (New CIA). In differing ways, both suggest the political possibilities inherent in myriad other initiatives linking the globally extensive, socio-ecologically complex processes of climate change with the non-climate concerns of established and emerging constituencies.

Synthesizing the material in Parts I and II analytically, Chapter 6 explores three key aspects of the politics of connection, drawing out their relevance for wider debates in scholarship and politics. The chapter and volume close by indicating some of the opportunities and dilemmas climate justice advocates and activists, and those who would ally with them, face going forward. Those conclusions extend from the relational analytical perspective on climate justice struggles developed and deployed throughout the volume. It is to that perspective, and how it elucidates both the production and contestation of climate injustice, that I now turn.

Bibliography

Arts, B. (1998). *The Political Influence of Global NGOs: Case Studies on the Climate and Biodiversity Conventions*. Utrecht: International Books.

Babbie, E. R. (2007). *The Practice of Social Research*. Belmont: Wadsworth Publishing Co.

BBC News. (2009, December 19). Copenhagen Deal Reaction in Quotes. *BBC*. Retrieved from http://news.bbc.co.uk/2/hi/science/nature/8421910.stm

Bernauer, T. (2013). Climate Change Politics. *Annual Review of Political Science, 16*(1), 130301143509009.

Betsill, M. M., & Corell, E. (2008). *Ngo Diplomacy: The Influence of Nongovernmental Organizations in International Environmental Negotiations*. Cambridge, MA: MIT Press.

Bond, P. (2011). *The Politics of Climate Justice*. London: Verso and Pietermaritzburg: University of KwaZulu-Natal Press.

Broder, J. M. (2011, December 10). In Glare of Climate Talks, Taking on Too Great a Task. *The New York Times*. Retrieved from http://www.nytimes.com/2011/12/11/science/earth/climate-change-expands-far-beyond-an-environmental-issue.html

Chatterton, P., Featherstone, D., & Routledge, P. (2013). Articulating Climate Justice in Copenhagen: Antagonism, the Commons, and Solidarity. *Antipode, 45*(3), 602–620. https://doi.org/10.1111/j.1467-8330.2012.01025.x.

Derman, B. B. (2014). Climate Governance, Justice, and Transnational Civil Society. *Climate Policy, 14*(1), 23–41.

Featherstone, D. (2008). *Resistance, Space and Political Identities: The Making of Counter-Global Networks*. Hoboken: Wiley-Blackwell.

Featherstone, D. (2012). *Solidarity: Hidden Histories and Geographies of Internationalism*. London: Zed Books.

Fisher, D. R. (2010). COP-15 in Copenhagen: How the Merging of Movements Left Civil Society Out in the Cold. *Global Environmental Politics, 10*(2), 11–17.

Gardiner, S. M., & Weisbach, D. A. (2016). *Debating Climate Ethics*. Oxford: Oxford University Press.

Harvey, D. (1996). *Justice, Nature and the Geography of Difference*. Hoboken: Wiley-Blackwell.

Johnson, C. (2012, January 26). *Denmark: Court Upholds Decision in Mass Arrests Case* [web page]. Retrieved March 18, 2015, from http://www.loc.gov/lawweb/servlet/lloc_news?disp3_l205402960_text

Klinsky, S., & Brankovic, J. (2018). *The Global Climate Regime and Transitional Justice*. London/New York: Routledge

McKie, R., & Zee, B. van der. (2009, December 12). *Hundreds Arrested in Copenhagen as Green Protest March Leads to Violence*. Retrieved March 18, 2015, from http://www.theguardian.com/environment/2009/dec/13/hundreds-arrested-in-copenhagen-violence

Newell, P. (2006). *Climate for Change: Non-State Actors and the Global Politics of the Greenhouse*. Cambridge: Cambridge University Press.

Paterson, M. (2011). Selling Carbon: From International Climate Regime to Global Carbon Market. In J. S. Dryzek, R. B. Norgaard, & D. Schlosberg (Eds.), *The Oxford Handbook of Climate Change and Society* (pp. 611–624). Oxford: Oxford University Press.

Peet, R., Robbins, P., & Watts, M. (2010). Global Nature. In R. Peet, P. Robbins, & M. Watts (Eds.), *Global Political Ecology* (pp. 1–48). New York: Routledge.

Roberts, J. T., & Parks, B. (2006). *A Climate of Injustice: Global Inequality, North-South Politics, and Climate Policy*. Cambridge: MIT Press.

Routledge, P. (2011). Translocal Climate Justice Solidarities. In J. S. Dryzek, R. B. Norgaard, & D. Schlosberg (Eds.), *The Oxford Handbook of Climate Change and Society* (pp. 384–398). Oxford: Oxford University Press.

Shearer, C. (2011). *Kivalina: A Climate Change Story*. Chicago: Haymarket Books.

Shift Magazine. (2010). Interview with Erik Swyngedouw: The Post-Politics of Climate Change, (8). Retrieved from http://www.indymedia.org.uk/en/2010/02/446191.html

Stephenson, W. (2015). *What We're Fighting for Now Is Each Other: Dispatches from the Front Lines of Climate Justice*. Boston: Beacon Press.

The Danish Institute for Human Rights. (2012, January). *Answers on Best Practices That Promote and Protect the Rights to Freedom of Peaceful Assembly and of Association*.

Retrieved from http://www.ohchr.org/Documents/Issues/FAssociation/Responses2012/
NHRI/Denmark.pdf

The Guardian. (2009, December 16). *In Pictures: Reclaim Power Climate Protest March in Copenhagen*. Retrieved March 18, 2015, from http://www.theguardian.com/environment/gallery/2009/dec/16/reclaim-power-march-copenhagen

Tokar, B. (2010). *Toward Climate Justice*. Porsgrunn: Communalism Press.

Vidal, J., Stratton, A., & Copenhagen, S. G. in. (2009). *Low Targets, Goals Dropped: Copenhagen Ends in Failure*. Retrieved February 10, 2015, from http://www.theguardian.com/environment/2009/dec/18/copenhagen-deal

Wallace-Wells, D. (2019). *The Uninhabitable Earth: Life After Warming*. New York: Crown/Archetype.

Zee, B. van der. (2010, December 16). *Danish Police Ordered to Compensate Climate Protesters*. Retrieved March 18, 2015, from http://www.theguardian.com/environment/2010/dec/16/danish-police-protesters-compensation

Producing and Contesting Climate Injustice

1

Climate injustice is produced by specific relationalities. Climate justice initiatives raise challenges to and endeavor to supplant those relationalities. This chapter considers some fundamental conditions of both processes, establishing as it does so the contours of their substantive origins as well as the analytical perspective that will animate subsequent chapters. That perspective combines the basic science of climate change and human dimensions thereof with a set of related empirical and conceptual concerns in geography, political ecology, socio-legal and social movement studies, and political theory. Together these help to elucidate both the challenges and the promises of climate justice as a political project.

Human Impact

Climate science plays a crucial role in recognizing and contesting climate injustice. Recent methodological developments make it possible, for instance, to express the link between many extreme weather events and the wider process of climate change, in terms of the probability of those events occurring under different scenarios of greenhouse gas emissions (see Chap. 2).[1] These "attribution studies," in turn, enable more precise assessments of the human and financial costs of climate

---

[1] See, for example, WRI 2013.

© The Author(s) 2020
B. B. Derman, *Struggles for Climate Justice*,
https://doi.org/10.1007/978-3-030-27965-3_1

change, both past and future.[2] In these and other ongoing scientific advances, greater precision and predictive power are made available by more data and better understanding of linkages between the multiple Earth system processes affected by, and affecting, the climate. Because those systems and processes themselves interact, and because still unknown threshold conditions may act as "tipping points" into "runaway" climate change or a rapid shift into a new, stable "hot Earth" state, crucial uncertainties remain. At the same time, as studies by different authors have accumulated, their agreement has been remarkable. Moreover, the basic processes by which human activity is changing the climate were reasonably well understood by the late nineteenth century.

As a result of human and non-human activities, molecules of carbon dioxide, methane, water vapor, and other greenhouse gases accumulate in the Earth's atmosphere. The branched molecular structures of these gases temporarily absorb, and then re-emit the energy of long-wave radiation leaving the Earth (originating as solar radiation). That energy can then persist in the atmosphere as heat, suggesting to early observers the effect of the glass roof of a horticultural greenhouse. Because of the global extent of atmospheric circulation, the greenhouse effect affects the entire globe. It is integral with life as we know it, in that living organisms release greenhouse gases, and without their presence in the atmosphere the nighttime cooling of the Earth's surface would produce overall temperatures far colder than those in which the complex ecologies of existing life evolved. Swedish physical chemist Svante Arrhenius described the greenhouse effect remarkably accurately in 1896, and predicted its intensification as an inevitable byproduct of increasing fossil fuel combustion associated with the industrial revolution.[3] That intensification is the primary cause of contemporary global warming and its heretofore unprecedented rate. Because of the persistence of greenhouse gases in the atmosphere, and the delays and threshold values that characterize the response of various environmental systems to the forcing of greenhouse-induced temperature and precipitation changes, the impacts of emissions on a given day influence climatic and Earth system responses over centuries. Anthropogenic emissions currently affecting change include contributions spanning dozens of generations, while today's emissions will exert effects as lasting.

Observations and predictions of climate change focus largely on differences in recorded or future values of temperature and precipitation, measured against baseline values at designated points in the past. Changes in these variables are linked

---

[2] For example, US Global Change Research Program 2018.
[3] Henson 2014.

with an extensive list of processes and events at a variety of timescales.[4] These include sea level rise, higher storm intensity, increased storm surge in coastal areas, changing spatial and temporal patterns of storms (including hurricanes and typhoons) as well as currents, glacial retreat and loss, melting sea ice, thawing permafrost and related subsidence, coastal erosion, ocean acidification, species decline (as well as loss and population explosion), shifts in the seasonality of rain and snow, decreased frequency and greater intensity of precipitation events and associated impacts on stream flow and inundation, mountain erosion from glacial retreat and stream flooding, and shifting biogeographies of plant and animal habitat and range.[5]

Many such effects are, in and of themselves, potentially disruptive of human life and livelihood. Indeed, the more rapid among them encompass much of the event space examined in human-environment research on hazards, and described frequently in legal language as "acts of God." Moreover, the deleterious effects of non-climate-related hazards can be dramatically intensified by rapid or slowly developing effects of climate change on relevant environmental systems.[6]

The lengthy list of direct and indirect impacts above also implicates myriad increasingly well-documented "downstream" effects associated with climate change, as these generate ripples throughout ecological webs as well as human economic, food, health, and infrastructural systems. Such climate-related socio-ecological impacts include crop failure resulting from higher growing season temperatures, less frequent and/or more intense rain events, and decreased availability of fresh water for irrigation from rivers, streams, and groundwater sources.[7] Unprecedented suicide rates among Indian farmers in recent years have been linked to the deleterious effects of such events on subsistence agricultural production.[8] New and intensified disease vectors are also arising, from combinations of changes in temperature,

---

[4] Causal links go both ways, creating positive feedback loops; for example, while sea level rise is caused by the melting of the ice caps at the Earth's poles due to warmer temperatures, the melting of permafrost also releases stored methane into the atmosphere, which in turn strengthens the greenhouse effect. The melting of ice and snow due to rising temperatures also changes the albedo—or reflectivity—of the ground surface, increasing the absorptive of energy at the Earth's surface and thereby accelerating melting.

[5] IPCC 2013.

[6] Consider, for instance, the greater potential devastation of coastal communities caused by tsunamis as the sea level rises.

[7] IPCC 2013.

[8] Renton 2011.

precipitation, and industrial agricultural practices.[9] The impacts of heat stress itself
on the human body threaten life, especially for the young, infirm, poor, and elderly.[10] In a 2014 estimate now considered conservative, the World Health
Organization (WHO) estimated climate change-related human deaths would reach
250,000 annually by 2030.[11] Some analysts have linked social conflict at the level
of whole nations and regions with the impacts of climate change.[12]

While human impacts are clearly the primary focus of such analyses, because the
climate system actually links all the physical and biotic Earth systems, they are each
affected by the anthropogenic intensification of the greenhouse effect. The magnitude
of these systemic changes and their increasingly apparent connection with human influence is reflected in increasingly ubiquitous references to "the Anthropocene," an
apparent geological epoch during which humanity's effects on the planet must be
considered not in the self-referential temporalities of human history, but the longer
time scale of Earth history itself.[13] That is, anthropogenic environmental changes unfolding today could well outlast the species that wrought them. The Anthropocene is
associated with multiple human effects, including what analysts have dubbed the
Sixth Mass Extinction, partially produced by climate change, wherein plant and animal species disappear at approximately 1000 times the background rate.[14] The scope
of Anthropocene human impacts is thus broader than climate change, but because it
influences biodiversity loss and other Anthropocene effects, and because the era is
often considered to have begun with the industrial revolution, anthropogenic climate
disruption could be described as its flagship process.

The Co-production of Global Climate Change and Socio-spatial Inequality

In one sense, climate change and the Anthropocene more broadly are obviously
global conditions. The socio-ecological processes that culminate in human impacts from anthropogenic climate change wreak havoc on conditions of social life

---

[9] These have been linked to new strains of pandemic influenza that move easily between species hosts and spread with human and commodity transportation, for instance. Wallace 2014.

[10] Record heat in Europe in the summer of 2003, for instance, is thought to have claimed as
many as 70,000 lives, with skewed impact on the elderly. Robine et al. 2008.

[11] WHO 2014; Haines and Ebi 2019.

[12] See, for example, Parenti 2012.

[13] Crutzen and Stoermer 2000; Chakrabarty 2009.

[14] See, for example, Kolbert 2014.

without regard to the territorial boundaries of nation-states or the originary locations of greenhouse gas emissions. They are "global" in the "border-crossing" sense.

At the same time, however, the impacts of climate change are by no means global in a second sense implying universality or evenness. In fact, the spatial disaggregation of climate change and its human causes and consequences uncovers some strikingly uneven geographies. Moreover, those geographies are patterned in ways that echo and re-inscribe centuries of uneven economic development since, to a large degree, the two are connected.

At a first cut, however, even the border-ignorant biogeochemical and physical mechanisms through which emissions generate distant impacts on the ground turn out to be far from global in the second sense. This unevenness results from the irregular continental and oceanic physical geography of the planet in combination with its shape, orientation, and motion. Despite atmospheric mixing, the Earth system impacts of increasing greenhouse gases affect different latitudes, surfaces, and specific locations quite differently (Fig. 1.1). The Arctic Climate Impact Assessment of 2005 and subsequent Intergovernmental Panel on Climate Change (IPCC) reports made clear that regions are not only impacted differently, but at widely varying rates.[15] The Arctic is warming significantly faster, for instance, than any other part of the globe.

---

[15] Arctic Council et al. 2005; IPCC 2013.

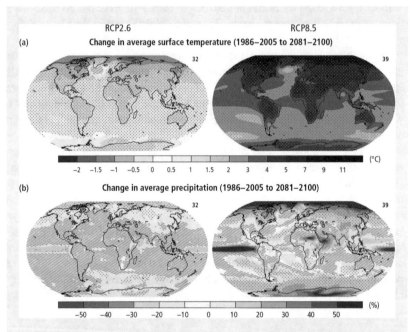

Fig. 1.1 Spatial variations in predicted temperature and precipitation change under two emissions scenarios, or Representative Concentration Pathways (RCPs), from the IPCC Fifth Assessment Report (AR5). The RCP2.6 scenario is modeled on recent proposals to limit average warming to 2 degrees Celsius. RCP8.5 represents a future without a specific mitigation target (Riahi et al. 2011). The maps compare (IPCC 2013, 12): "Change in average surface temperature (a) and change in average precipitation (b) based on multi-model mean projections for 2081–2100 relative to 1986–2005 under the RCP2.6 (left) and RCP8.5 (right) scenarios. The number of models used to calculate the multi-model mean is indicated in the upper right corner of each panel. Stippling (i.e. dots) shows regions where the projected change is large compared to natural internal variability and where at least 90% of models agree on the sign of change. Hatching (i.e. diagonal lines) shows regions where the projected change is less than one standard deviation of the natural internal variability."

At a second cut, such spatially variant physical impacts affect regionally and locally differentiated biological systems. Resulting changes in species distribution and abundance will continue to impact food supplies, such as fish catch, in different ways around the globe (Fig. 1.2).

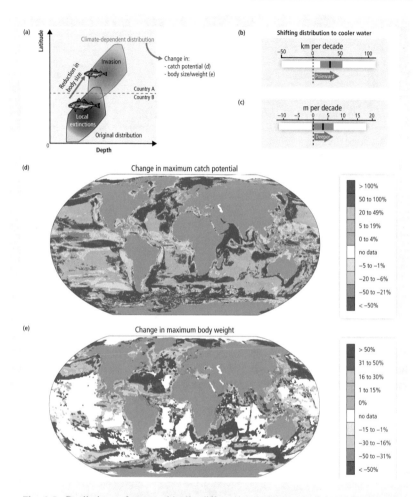

Fig. 1.2 Predictions of geographically differentiated changes in fish catch potential from IPCC AR5. The predictions are based on scenarios from previous reports, with emissions somewhat below those of AR5's RCP8.5 (IPCC 2013, 458): "Climate change effects on the biogeography, body size, and fisheries' catch potential of marine fishes and invertebrates. (a) Shifts in distribution range and reduction in body size of exploited fish driven by projected warming, oxygen depletion, and sea ice retreat (cf. [IPCC 2013] Figs. 6–7). Whenever the shift in distribution does not fully compensate for warming and hypoxia, the result will be a decrease in body size. Shifts in (b) latitudinal and (c) depth distribution of 610 exploited demersal fishes are projected to have a median

(continued)

(central line of the box) of 31 km per decade and 3.3 m per decade, respectively, with variation between species (box boundary: 25th and 75th percentiles) from 1991–2010 to 2041–2060 under the SRES A2 (between RCP6.0 and 8.5) scenario (Cheung et al. 2011, 2013). (d) Combining species' range shifts with projected changes in net primary production leads to a projected global redistribution of maximum catch potential. (Analysis includes approximately 1000 species of exploited fishes and invertebrates, under warming by 2 °C according to SRES A1B ($\approx$RCP6.0), comparing the ten-year averages 2001–2010 and 2051–2060; redrawn from Cheung et al., 2010). (e) Changes in species distribution and individual growth are projected to lead to reduced maximum body size of fish communities at a certain site. The analysis includes 610 species of marine fishes, from 1991–2010 to 2041–2060 under SRES A2 (approximately RCP6.0 to 8.5; Cheung et al. 2013), without analysis of potential impacts of overfishing or ocean acidification. Key assumptions of the projections are that current distribution ranges reflect the preferences and tolerances of species for temperature and other environmental conditions and that these preferences and tolerances do not change over time. Catch potential is determined by species range and net primary production. Growth and maximum body size of fishes are a function of temperature and ambient oxygen level."

Moving further through the socio-ecological linkages upon which human lives and livelihoods depend introduces additional layers of complexity. Here too, however, research indicates that climate change affects what the IPCC calls "human and managed systems"—which include food production, health, livelihoods, and economies—in geographically uneven ways (Fig. 1.3).

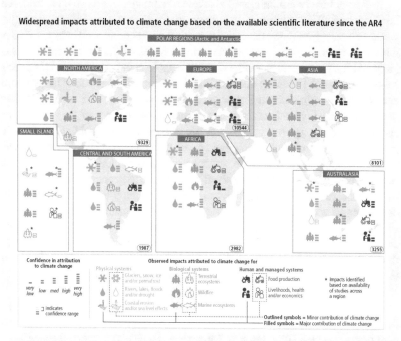

Fig. 1.3 Widespread impacts of climate change as summarized in AR5 by the IPCC working group on impacts, adaptation, and vulnerability (WGII) (IPCC 2013, 7): "Based on the available scientific literature since the IPCC Fourth Assessment Report (AR4), there are substantially more impacts in recent decades now attributed to climate change. Attribution requires defined scientific evidence on the role of climate change. Absence from the map of additional impacts attributed to climate change does not imply that such impacts have not occurred. The publications supporting attributed impacts reflect a growing knowledge base, but publications are still limited for many regions, systems, and processes, highlighting gaps in data and studies. Symbols indicate categories of attributed impacts, the relative contribution of climate change (major or minor) to the observed impact, and confidence in attribution. Each symbol refers to one or more entries in WGII Table SPM.A1, grouping related regional-scale impacts. Numbers in ovals indicate regional totals of climate change publications from 2001 to 2010, based on the Scopus bibliographic database for publications in English with individual countries mentioned in title, abstract, or key words (as of July 2011). These numbers provide an overall measure of the available scientific literature on climate change across regions; they do not indicate the number of publications supporting attribution of climate change impacts in each region. Studies for polar regions and small islands are grouped with neighboring continental regions. The inclusion of publications for assessment of attribution followed IPCC scientific evidence criteria defined in WGII Chapter 18. Publications considered in the attribution analyses come from a broader range of literature assessed in the WGII AR5. See WGII Table SPM.A1 for descriptions of the attributed impacts."

If geographically disaggregated analyses of climate impacts suggest the likelihood of dramatically differing consequences for human populations around the world, critically examining the socio-ecological connections involved in producing those consequences casts them in stark terms of disproportionate responsibility and vulnerability. The recent scientific and journalistic embrace of the Anthropocene concept indexes the growing magnitude and widening recognition of human influence over the Earth's physical processes. Social scientists were already primed, though, to critically analyze those developments, by decades of socio-ecological enquiry. Here too, uneven geographies are pervasive. Early political ecology, in particular, often centered the intertwining of the ecological and the social, tying environmental change to social processes in "chains of causation" across spatial scales. These analyses often explained human-induced ecosystem deterioration, for instance, as the result of local social marginalization produced in tandem with distant capital accumulation.[16] From these and other types of studies, it is clear that social power is directly involved in "making the world," as Erik Swyngedouw called such constitutive socio-ecological connections. Social power operates through, and often depends upon, the control of ecologies, leading to environmental change and destruction as well as new forms of social control, conflict, and resistance.[17]

Seen in this light, Earth in the Anthropocene must also be understood as produced through specific social and socio-ecological relations, shot through with both power and inequality. Jason W. Moore, for instance, reframes the epoch as the Capitalocene, in which the planet has been profoundly shaped by capital's need for cheap labor, energy, food, and raw materials.[18] Donna Haraway et al. cast the dominance of landscape configurations involved in procuring those requirements as the Plantationocene, a boundary event, in geological perspective, at which vast amounts of land and water are taken up in industrial farming, to the detriment of biodiversity and climate, as well as balance and affinity in the human orientation toward the non-human.[19]

Arguably, however, emphasizing in such terms the social production of climate change and other global system disruptions still risks reifying a presumed ontological division between "nature" and "society." Bruno Latour argued that

---

[16] See, for example, Blaikie and Brookfield 1987.
[17] See, for example, Robbins 2004; Rutherford 2007.
[18] Moore 2015.
[19] Haraway 2015; Haraway et al. 2016.

this distorted perception, of what is in fact a world of more-than-human hybrids, or assemblages, shaped western thought in particular since the enlightenment, in a hegemonic "modern constitution."[20] That is, institutionalized and reified in language, the presumption of nature-society division and the epistemological habits that flow from it can foreclose understandings and political possibilities that are urgently needed in the face of "monsters" unleashed by more-than-human, techno-scientific practices like a fossil fuel-based energy system and its disruption of the climate. Exposing the occlusions of the modern constitution's insistence on separation, materially-oriented, socio-technical analyses have sought to highlight the roles that non-humans (technologies and ideas, as well as chunks of "nature") play as participants in more-than-human assemblages, forcefully shaping human histories and politics. Timothy Mitchell shows, for instance, that as human societies came to rely on energy from petroleum, its material characteristics came to exert profound influence over political possibilities as well as climatic futures.[21]

Political possibilities and socio-ecological futures are also, however, produced by historical and geographical circumstances, and struggles founded upon them (Chap. 6 argues for this reading more fully). Indeed, a geographically relational engagement with critiques like those of Moore, Haraway, and Mitchell suggests, as do several climate justice groups, that global climate change must be understood as co-produced with spatial and social inequality, via socio-ecological ties and processes. The most readily recognizable index of this co-production is probably the association between greenhouse gas emissions and indicators of social and economic development, apparent from data available at the level of the nation-state. Human Development Index (HDI), for instance, rises in roughly direct proportion with exponentially increasing per capita emissions, since the basket of development indicators it measures were achieved and are maintained in large part through the expenditure of fossil fuel energy (Fig. 1.4).[22]

---

[20] Latour 1993.
[21] Mitchell 2011.
[22] UNEP 2014.

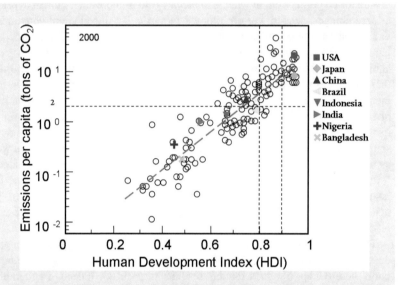

Fig. 1.4 Correlations between HDI and per capita CO_2 emissions for the year 2000 (Costa et al. 2011). The authors note: "The dashed line represents a least squares fit through all values. The coefficient of determination is $R^2 \simeq 0.8l$ and the correlation coefficient is $\rho \simeq 0.90$. For some countries the values are shown explicitly. Vertical lines represent the HDI values of 0.8 and 0.9, representative of high and very high development standards, respectively, as expressed in the United Nations Development Report 2009. The horizontal line shows the 2 tons per capita CO_2 emissions target to limit global warming at 2 °C by 2050 (UNDP 2009; WBGU 2009)."

Unsurprisingly, these lead to an increased ability of human populations to weather, as it were, both sudden and gradual impacts from climate change. Conversely, lack of "development" produces vulnerability to those impacts. That is, the character, severity, and distribution of the *re*-socialization of human-induced, climate-related environmental change depends to a significant degree on already-existing social conditions.[23] The concept of vulnerability and its inversion in resilience or adaptive capacity are therefore cornerstones of the human impacts literature. Figure 1.5, included in the IPCC Fifth Assessment Report, for instance, illustrates how risks associated with climate change result from combinations of climate system impacts (including those driven by humans) and socio-economic drivers of vulnerability and exposure.

---

[23] Their anthropogenic production aside, that is, climate change impacts, like other hazards, is socio-natural, rather than "natural" in their effects. Hewitt 1983; Shearer 2011.

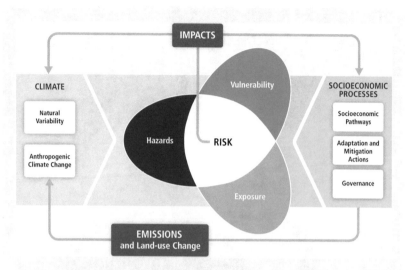

Fig. 1.5 The IPCC's 2013 model of risk (IPCC 2013, 3): "Illustration of the core concepts of the WGII AR5. Risk of climate-related impacts results from the interaction of climate-related hazards (including hazardous events and trends) with the vulnerability and exposure of human and natural systems. Changes in both the climate system (left) and socio-economic processes including adaptation and mitigation (right) are drivers of hazards, exposure, and vulnerability."

Like the atmospheric and biophysical parameters of climate impact, these socio-economic pathways vary spatially in significant ways. Moreover, because of the co-production of global climate change and regionalized "human development," the geographies of responsibility for and vulnerability to climate-related impacts end up as rough inversions of each other. Remaining, for the moment, at the analytical scale facilitated by national data, world cartograms deforming nation-state areal units in proportion to selected non-spatial variables ably illustrate some of these unsettling disproportionalities (see Fig. 1.6). The WHO's region-level estimates of deaths from climate change confirm that climate-related mortality, while globally extensive, is far more common in countries with lower scores on development metrics.[24]

---

[24]WHO 2007; see country-level map of incidence at WHO 2019.

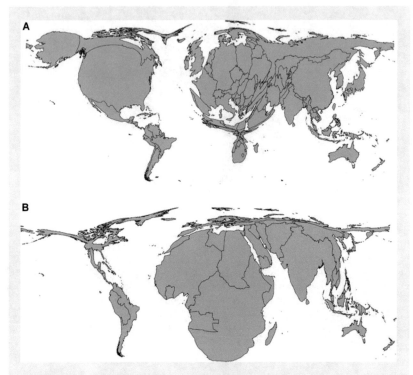

Fig. 1.6 Patz et al.'s comparison of "[d]ata-driven cartogram maps demonstrating (**A**) relative proportions of cumulative CO_2 emissions, by country, and (**B**) magnitude and severity of the consequences of climate change for malaria, malnutrition, diarrhea, and drownings, by country (Patz et al. 2007)."

Importantly, however, such comparisons obscure the constitutive connections underpinning different levels of national development, as well as the deep sub- and transnational disproportionalities that also characterize climate-related harm and responsibility for its production. To expand on the first point, comparisons of attributes compiled at the nation-state level construct those states analytically as functionally separate territorial containers within which events occur, apparently, independently.[25] Relational analyses, instead, highlight the ways in which such

---

[25] Agnew 1994.

places are in fact co-produced, through globally extensive social and material ties.[26] Geographically uneven patterns of development, for instance, follow in large part from the operation of specific forms of connection *between* places and *across* space. Economic boons and their social benefits accrue in certain places in conjunction with exploitation and extraction in others, as part of the global capitalist economy's need to expand value, and the spatial extensions and reconfigurations that have facilitated that requirement.[27]

Re-examining the territorial assignment of greenhouse gas emissions with such relational considerations in mind helpfully illustrates some of the ways in which the apparent separation of places, people, and decisions in different nation-states obscures their mutually constitutive, socio-ecological intertwining. Luke Bergmann's analyses of the commodity and capital flows associated with 2004 emissions, for instance, undercut the assessment of responsibility for climate change in terms of yearly national totals, a measure by which China has recently lead, since a significant portion of its emissions were linked with consumption in the historically high-emitting US and Europe. Accounting for, and mapping, the association of those emissions with capital accumulation offers an even starker depiction of the long-industrialized countries' continuing disproportionate responsibility for anthropogenic climate change, in spite of the recent preponderance of yearly totals (and their immediate pollution impacts) in China (Fig. 1.7).[28]

---

[26] The literature on relational geographies is vast; touchstones of the proceeding analysis include Harvey 1982; Massey 2005a; Lawson 2007; Smith 2008.

[27] See, in particular, Harvey 1982.

[28] See also Bergmann 2013. These dynamics have shifted somewhat in subsequent years as consumption and accumulation has grown in China, and as it has itself come to draw extractively on labor and resources in other countries. See, for example, Harvey 2019. The point remains, however, that assigning emissions to territory can obscure the locations of their driving conditions and benefits.

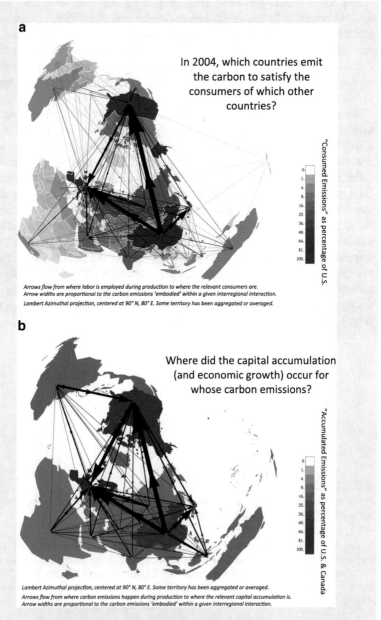

Fig. 1.7 Maps of interconnection between countries through carbon emissions associated with extraterritorial consumption (**a**) and capital accumulation (**b**) (Bergmann 2011, used with permission. See also Bergmann 2013)

Telling as these snapshots of socio-ecological connection between countries are, the sub- and transnational disproportionalities and historical processes they elide are also crucial to a multi-dimensional, relational conception of climate injustice.

Time, Space, Power

Connections and inequalities between contemporary nation-states reflect a long history of globalization—linking multiple episodes of resource and human plunder; colonial exploitation, occupation, and settlement; military and economic imperialism; and extractive forms of trade and aid—throughout which both wealth and emissions accumulated in the first communities to industrialize with the help of coal and, later, oil. A series of specific governmental relations developed alongside these differentiating forms of global connection, often serving to enshrine or facilitate uneven geographies of social power and entitlement to "nature." These formalizations of relations between groups and places—and the periodic exceptions during which their stabilizing rules have been laid aside—have often drawn upon or underwritten controlling and exclusionary conceptions of social and spatial difference, couched in terms of racial and ethnic superiority, nationalism, or security.[29]

At the level of the globe, the modern nation-state and the collection of laws and institutions assembled around it provide core governmental structures: containing, exposing, and to some extent mediating inequality between populations. Crucially, the Westphalian principles that shape this basic framework of international relations presumed the legitimacy and enshrined the sovereign independence of separate, territorially defined polities. Multiple international human rights and environmental treaties, for instance, articulate the outlines of shared commitment by states to their citizens or to each other, but they also reflect in legal terms their political and social isolation. Legal human rights instruments therefore offer only limited protection against the abuse of power within countries or the border-crossing effects of economic and ecological harm, including climate change (see Chap. 2).[30]

International legal rules can also be understood in terms of the levers, protections, and limits they furnish for supranational economic projects. From the earliest

---

[29] Harvey 1982; Blaut 1993; Dalby 2002, 2009; Peet 2003; Smith 2003; Mitchell 2011.
[30] Osofsky 2006.

rules formalizing practices of extraterritorial exchange, to latter twentieth-century trade agreements and intellectual property safeguards, to the evolving governance of financial flows, law has developed together with economic globalization in ways that largely facilitated the concentration of capital in industrial and banking centers.[31] To the extent those frameworks and the institutions that administer them come to structure climate change governance, therefore, they also open pathways by which economic inequalities can be re-inscribed as an adjunct of climate action.[32]

Indeed, a wide range of forums convene around the administration of international legal frameworks and substantive issues of global concern, with potential relevance for coordinated responses to climate change. These vary in official and advisory capacities, more or less forcefully mediating relations among nation-states and transnational social groups. To a large degree, international organizations exist to ameliorate the worst outcomes of relationally produced global inequality and environmental deterioration, through dialogue, cooperation, or aid. In a wider instrumental sense, that work helps to stabilize the existing international order, against threats such at armed conflict. In so doing, it has also institutionalized relations of unequal power, sometimes by design.[33] In their deliberative procedures, for instance, international institutions offer venues for the pursuit of nationally defined economic and social interests, as well as those of a cosmopolitan elite empowered by their ability to participate independently and as, or through, national delegates. Conversely, international institutions and laws can relegate Indigenous peoples and other groups without state status or influence to the margins of governance, and beyond official mechanisms of recourse.[34] International forums can also simply provide a playfield for the power differences between states. Although the UNFCCC, for instance, exists specifically to address climate change, and though its founding documents acknowledge crucial differences in responsibility and capacity between nation-states, it also institutionalizes their status as equal negotiating parties in consensus decision making, formally empowering the minority of wealthy, historically high-emitting industrialized countries which are privileged in international relations by default (see Chap. 3).[35] Finally, claims on exclusive access to fossil fuel resources can also become a stake in the design and

[31] Gill 1995; Sparke 2012.

[32] Lohmann 2008.

[33] Smith 2003; Peet 2003; Mitchell 2011.

[34] Osofsky 2006.

[35] Gardiner 2011.

conduct of international forums, and in military ventures with the blessing of international or multilateral accord.[36]

Governmental relations at the international level, then, play an important role in relation to globally uneven social geographies, just as their underlying principles can obscure the more-than-human histories of connection between places and peoples that ground those contemporary spatial inequalities. Arguably, the institutional framing of international relations as negotiation, cooperation, or assistance among territorial nation-states in turn helps to reinforce a particular conception of development, with its corresponding relations and practices. (Inasmuch as the UNFCCC differentiates the responsibilities and capacities of negotiating parties, for example, it does so through language designating them as "developed" or "developing.") This conception of development envisions countries as situated along a trajectory traced first by the high-emitting industrialized countries and marked out, typically, in metrics of economic growth that remain unmoored—thanks to a formative period of abundant fossil energy and de-materialized financial flows—from the ecological foundations it now threatens.[37]

Of course, differences in "development" achieved through fossil-fueled economic activity and measurable by monetary indexes like Gross Domestic Product (GDP) usually do dramatically differentiate the climate-related risks populations face. Spatialized governmental separations and social statuses help render these differences even more starkly and disparately. That is, national (and other) boundaries are in fact only *partially* permeable with respect to the human impacts of global climate change. Borders and citizenship both engender significant variation in the material consequences of impacts and response measures, given the vastly unequal resources available for resilience and climate change adaptation in different nations and locales. Examples include rights of international passage enjoyed by some individuals but not others, global health initiatives constraining climate-related disease threats to particular "hot spots," and the disparate financial capacities of states to construct relatively simple functional seawalls.[38]

Such consequential differences in climate-related risk between populations (and their approximate inversion of responsibility for climate-related harm) also hold within states, and between social status groups that span borders. As Stephen Gardiner notes, there is "a Germany in China ... [and] something like a Pakistan or

---

[36] Mitchell 2011.

[37] Blaut 1993; Escobar 1995; Mitchell 2011.

[38] See, for example, Sparke 2005, 2006; Dalby 2002, 2009.

Bangladesh too."[39] Such subnational differentiations are also produced relationally, through specific forms of connection between groups and places, both within and across national borders. Doreen Massey shows, for instance, how London's emergence as one of the pinnacles of financial and cultural globalization corresponded with the deepening of economic and social divisions between southern and northern England, as working people were priced out of housing in the national/international center, which became the home, instead, of a cosmopolitan elite that both commands greater financial resources and expends, on average, more energy.[40]

Socio-spatial inequities of climate-related risk and responsibility within (and across) borders are also tied to racialized violence and oppression, with their own national and global histories. Poor and historically marginalized groups live in disproportionately close proximity to the facilities where fossil fuels are produced, and are therefore disproportionately subject to the very local impacts on health and human development associated with workaday activity at those sites of production. The "spillover effects" of higher than average exposure these groups typically experience from polluting land uses at the upstream sources of climate-changing emissions are a significant subset of well-documented and theorized environmental racism/injustice in the US and other settings.[41]

Climate change also creates the conditions for a kind of spatio-temporal telescoping of risk in marginalized communities, however: co-locating the disproportionate effects of longer-term Earth system processes with the short-term, local impacts of polluting nodes in the fossil energy system. This occurs wherever communities adjacent to production facilities are also differentially affected by the climatic consequences of emissions, wherever they occur. Such a double dose is only at its most apparent when, for instance, poor black communities on the Gulf Coast of the US or in the refining neighborhoods around the port of Durban are threatened by more frequently occurring high-intensity storms and sea level rise.

In fact, climate-related disproportionalities appear across many power gradients. Accumulating research suggests that the poor, young, elderly, non-white, female, LGBTQ (lesbian, gay, bisexual, transgender, and queer/questioning), and Indigenous groups are all more differentially vulnerable to climate change impact, and that these groups also often emit less than their more socially dominant counterparts. The spatiality of disproportionality is multi-dimensional as well, in that

---

[39] Gardiner 2011, 320.

[40] Massey 2005b; Oxfam 2015.

[41] See, for example, Harvey 1996; Lazarus 2004; Schlosberg 2007; Bullard 2008; NAACP 2012.

subnational, international, and transnational inversions of responsibility and risk can each be identified.[42]

Redressing such deeply rooted, socio-ecological inequalities has become increasingly difficult over the latter twentieth and early twenty-first century amidst shifts within many polities and in global governance toward market-based arrangements for the provision and allocation of social goods. Indeed, while market mechanisms for efficiency generally distribute benefit inequitably, in many settings greenhouse gas emissions and their detrimental consequences (including upstream local pollutants) also retain the status of unaccounted for "externalities." Moreover, market-based and market-oriented theories of governance have underwritten the widespread withdrawal of states (at a variety of administrative levels) from public goods provision, as well as the privatization of once publicly held resources and publicly administered services, followed, often, by reductions in state funding.[43] Resulting neoliberal relations of control, administration, and provision have shifted responsibility for crucial "socio-economic processes," to use the IPCC's risk model terminology, from government to market and/or civil society actors, with concomitant reductions in public sector capacity, democratic control, and space for political contest, while increasing inequality and largely ignoring climate-related crises.

Moreover, in reducing the funding and priority of longer-term planning and monitoring activities, the ascension of market-based and privately administered governance only exacerbates the mismatch between ecological time scales and political and economic ones. The gradual, incremental, and systemic qualities that characterize the many slowly developing processes of climate-related social harm contrast with the shorter cycles of electoral politics and corporate accounting, rendering them less likely objects of intervention backed by adequate financial or political investment.[44] Private and corporate influence over knowledge production and information provision also erode public confidence in knowledge of climate and related socio-ecological change. Privately funded studies casting climate knowledge in doubt fueled skepticism, confusion, and partisan politicization among wide swaths of the voting public in the US, for instance.[45] More generally, though, it simply runs counter to the imperative of continued accumulation in a fossil-fueled, consumption-driven capitalist economy to highlight the socio-ecological ties that threaten its continuity, particularly over longer time

---

[42] See, for example, Terry 2009; Wildcat 2013; Oxfam 2015; Jafry 2018.
[43] Harvey 2007.
[44] Nixon 2011.
[45] Layzer 2015.

horizons. There is a pronounced tendency, therefore, particularly in the "developed" world to overlook those ties as far as possible. Finally, fundamental cognitive factors apparently exacerbate these same tendencies toward inaction. The immense spatial and temporal scales of climate and other Earth systemic processes themselves discourage perception of connection between the daily conduct of fossil-fueled human life and its long-term, long-range ecological underpinnings and consequences.[46]

These relational factors each contribute to what ethicist Stephen Gardiner characterizes as a "perfect moral storm." In global, international, ecological, and intergenerational domains, Gardiner notes, those least responsible for climate change are (or will be) most harmed by its impacts. Moreover, because power and security tend to co-locate socially and spatially with emissions, those most responsible are also most able to respond, but least motivated to do so by their own conditions. Finally, the primary institutional settings available for the construction of coordinated responses tend to foster high emitters' dominance in decision making.[47]

Relational perspectives lay bare the co-production of these disproportionalities with climate change itself, through processes linking "nature" and "society," connections between peoples across time and space, and institutional structures that emerged from and now re-inscribe inverted geographies of responsibility and vulnerability, while obscuring their co-constitutive origins. As the activist quoted in the Introduction suggested, therefore, making climate justice depends on making connections across a host of domains. Like the processes driving it, that is, the terrain, strategies, and stakes in struggles against climate injustice are essentially relational. Combining analysis with practice, those strategies involve disaggregating, socializing, and historicizing the apparent globality of climate change; exposing its uneven geographies and their socio-ecological and -spatial underpinnings; and challenging governmental and social divisions and disempowerments in the name of more equal and sustainable futures.

Contesting Climate Injustice

Climate justice struggles are also shaped by advocates' and activists' particular experiences, resources, and ideological commitments, as well as their sensibilities to shifting and enduring conditions of opportunity and constraint in a given

---

[46] Lazarus 2004.
[47] Gardiner 2006, 2011.

context.[48] That such differentiating factors informed the progression of efforts examined in this volume suggests their relevance for other projects of climate-related protection, redress, or transformation as conceived and pursued by other actors, in other settings. At the same time, several conditions that cut across the efforts examined here also arguably shape possibilities and liabilities in the pursuit of climate justice more broadly. In identifying and characterizing these conditions, critical relational perspectives again furnish analytical insight.

Law provides language and forums for climate justice efforts and, to varying degrees, the possibility of material redress through the exercise of state power on behalf of disenfranchised and minority claimants.[49] As Gloppen and St. Clair describe in their own, parallel terms, each of the relational dimensions involved in producing climatic injustice also appear as crucial facets in the framing or outcome of a growing number of legal claims tied to climate-related harm.[50] Chapter 2 shows that, at the global level particularly, the complexity in relations of climate-related harm and disproportionality can pose significant obstacles to the recognition of responsibility through legal reasoning, which tends to reify the socio-ecological divides of Latour's "modern constitution," and to re-inscribe socially or politically definitive spatial divisions like those of the Westphalian compact.[51] Moreover, despite the rhetorical prominence of reason in legal rulemaking and interpretation, both provide openings for the play of uneven power relations, just as the venues and efficacy of legal judgment require the political will and material capacity of states. Further, as critical legal geographers have noted, law's role in helping create worlds in which "justice" and order proceed as fore-ordained depends crucially on its arbiters' control of space.[52] Border-crossing, more-than-human harms linked closely with the amassing and consolidation of geo-economic power, therefore, pose serious challenges for legalities focused on limits or responsibility.

Principles and procedures with legal status also provide structure to the unfolding of policy processes, which can themselves be complex as well as contentious. The invocation of core principles and the triggering or revision of key procedures, therefore, offer additional mechanisms for advocates seeking change: in this case

---

[48] The latter corresponds to "political opportunity structures" as theorized by social movement scholars. Meyer and Minkoff 2004. With particular relevance for the foregoing case analyses, see Miller 1994; Kriesi 1995; Van Der Heijden 2006.

[49] On legal mobilization more broadly, see Turk 1976; Zemans 1983; McCann 1994.

[50] Gloppen and St. Clair 2012.

[51] See also Delaney 2003, 2010.

[52] Blomley 1994; Blomley et al. 2001.

through the creation, implementation, or adjustment of policy, rather than the assessment of responsibility for harm. As in many settings, such structuring legalities have been the focus of prodigious advocacy in the UNFCCC, as states and nongovernmental observer groups pursue a variety of aims. Persistent legalistic principles and procedures do much to legitimate policy change, but their permanence also masks the ongoing potential for redefinition and interpretation under the influence, and according to the interests, of the more powerful participants in policy debates. The architecture of the climate convention evolved in this manner, between 2009 and 2016 (see Chap. 3), as relations between countries were fundamentally re-envisioned, their responsibilities for climate change and climate action re-assessed, and the relevance of legality itself in international responses cast in doubt.

It is, in fact, the inadequacy of law and policy in the face of global social and environmental problems as well as the shift from state-centered government to privately administered governance that has afforded non-governmental groups and other organizational forms of "civil society" greater recognition and deference in decision-making forums as well as scholarship.[53] Unsurprisingly, such groups have played central roles in analysis of and advocacy for social justice in the context of climate change (see, especially, Chaps. 4 and 5). Since at least 2000, when social movement, NGO, and community representatives gathered during COP6 in The Hague, their critical responses to the international climate regime have often been framed in terms of "climate justice."[54] Those impulses have expanded in geographic reach, topical scope, and salience over proceeding years, crystalizing periodically in the emergence and statements of leading groups and coalitions. Such episodes include the founding of the Durban Group for Climate Justice in 2004, Climate Justice Now! (CJN!) during the Bali COP in 2007, and Climate Justice Action (CJA) in the months preceding COP15.[55] These and other coalitions assemble a wide range of forces in civil society, including longstanding networks and organizations like Friends of the Earth International (FOEI), Indigenous Environmental Network (IEN), and La Via Campesina; other groups advocating debt forgiveness, development, human rights, or forms of food, gender, social, or environmental justice; representatives of labor, Indigenous peoples,

---

[53] Keck and Sikkink 1998, 1999; Salamon and Sokolowski 2004; Salamon et al. 2003; Kaldor et al. 2003, 2012.

[54] See, for example, CorpWatch 2000.

[55] See Newell 2005; Pettit 2004; Bond 2011, for discussion of the emergence of climate justice politics at the international level.

fisherfolk, and forest-dwellers, supporters of biodiversity and alter-globalization; progressive and radical think tanks; faith-based campaigners; thematic watchdogging groups; entities formed around "climate justice" itself; and disenchanted members of mainstream NGOs.[56]

If these climate justice-oriented groups formed crucial alliances and stances during global gatherings in and around COPs, other scales of political action gained relevance for many climate activists and policy makers as the Kyoto Protocol's disappointing mitigation outcomes became clear, and with the more precipitous decline in international ambition and accountability signaled in Copenhagen.[57] Indeed, post-Kyoto evolutions of the international climate regime underscored (and arguably increased) the influence of national interests, particularly those of the most responsible—and response-*able*—long-industrialized countries, and to lesser, varying degrees the rapidly industrializing countries and major oil-producing states.

The first of these country groups (including the US and European Union [EU] members) have provided the logical focus for much climate justice advocacy at UNFCCC meetings. Transforming the conceptions of national interest that direct negotiating positions and domestic policies with respect to climate change in these countries, however, means expanding and fragmenting the work of advocacy, since political opportunities differ both between polities and over time. As such, advocacy across multiple regimes is also difficult to coordinate.[58] Further, the relative insulation of many groups in the most responsible/response-able countries from climate-related harm to date means that its salience there has, for the most part, remained low in contrast to the influence of fossil fuel interests. Impelling adequate policy change in these settings could therefore require energetic and concerted political action by more insulated domestic subpopulations as well as disproportionately impacted ones and, in some places at least, those more vulnerable to the prospective impacts of transition to low carbon energy. Even within polities, the cultural characteristics, socio-economic conditions, and education levels of such groups can differ dramatically. Mobilizing them and other potential constituencies for climate action and/or justice therefore requires first rendering concerns about climate change (and, ideally, the injustices connected with it) in ways that resonate meaningfully for multiple, distinct social groups and settings. Chapters 5 and 6 examine recent efforts in this vein.

---

[56] The November 2010 list of Climate Justice Now! network members names nearly 800 such groups. See Climate Justice Now! 2010.

[57] See, for example, Routledge 2011.

[58] Derman 2014; see also Dryzek et al. 2003; Mahoney 2008.

Such politically activating communicative work lies at the foundation of broad-based transformative social change, but it need not, it bears underscoring, proceed apace or unproblematically.[59] Arguably, environmental problems, and climate change especially, have taken on a "post-political" character, such that the agonism of differing subject positions is sidelined in favor of aggregate emissions reductions however achieved, since they are understood to be collectively imperative.[60] In the context of neoliberal governmental relations, those mechanisms are likely to be constructed around continued private accumulation, amidst decreased public power and participation. To the extent non-governmental organizations assume representative as well as technocratic roles, and sometimes combine them in climate-related decision-making settings, those roles potentially implicate them in such depoliticizing processes. For instance, to the degree specialized technical, legal, or policy knowledge coupled with social privilege enable "representatives of civil society" to participate in international climate talks, their abilities to give voice to marginalized, disengaged, and disproportionally affected groups appears doubtful, as does the likelihood of their providing those groups with politically empowering reporting or analysis. Although clearly varied, national and subnational decision-making forums devoted to climate change (and other issues) can replicate such contradictions.

In fact, the incorporation of "civil society" in governance and policy-making processes is easily misidentified with broader promises of social benefit liberal political theories have attached to the term. Edwards identifies three distinct, though often conflated, meanings of civil society common in recent usage, which are helpful in disentangling those misidentifications. The inclusion of "civil society organizations" as participants or observers to decision-making processes corresponds to a first meaning of the term, referring to a specific organizational *form*: a type of social collective distinguishable from the state and, for most recent theorists, the market. Civil society can also, though, connote a set of *norms*: of conduct and communication associated with desirable conditions of life within a wider social body or polity. Finally, the term can suggest the existence of evenly accessible forums of social intercourse within which such propitious norms hold sway. Such a "public sphere" provides the discursive terrain for reasoned debate, inclusive of many voices and views, thought to undergird effective democracy.[61]

---

[59] Gramsci 1971; Hart 2006.

[60] Swyngedouw 2007, 2010.

[61] It is worth noting that Edwards' (2009) typology of meanings does not exhaust those associated with "civil society," particularly as deployed across successive periods of social theory. In Hegel's usage, for instance, the term is associated with the sphere of market competition, to the extent that Hegel's analysis of social conflict within "civil society" resembles

Clearly, although distinct, these different meanings of civil society are also normatively intertwined and practically connected. Indeed, as the proceeding chapters suggest, each is invoked or implied, in various times and places, in connection with the politics of climate justice. Moreover, activists and advocates themselves often deploy the term "civil society" in connection with the collectives and political activities of which they are a part, particularly around COPs and simultaneous "peoples" summits. (I therefore use the term advisedly but frequently as well, as I describe and analyze their work, and return to wider theoretical debates surrounding its political valence in Chap. 6.)

Still further, the idea of civil or "active" society plays key roles in certain transformative political theories. In Gramsci's analysis, the civil society of Italy in the early twentieth century (encompassing institutions and collectives of many types) provided crucial ideological support to a national political order veering into fascism, fortifying the latter against his own socialist revolutionary project. Rather than essentially counter-revolutionary, however, for Gramsci the "complexes of associations" in civil society were among the very "trenches" and "fortifications" in the terrain of political struggle.[62] In a similar, if inverted, manner, Polanyi's sweeping economic history presents "active" society as the locus of "countermovement" against the destructive commodification of land, labor, and money, previously stabilized by their embedding in social relations.[63] Importantly, Fraser builds on Polanyi's "double movement" of market and society, adding a third vector of "emancipation" in acknowledgment of the potential for domination *within* stable social orders as well as countermovement projects.[64]

Conceived inclusively in terms of form, then, to encompass not only professionalized NGOs of varying orientations but also less institutionalized collectives identified with established and emerging social movements,[65] "civil society" suggests a heterogeneous set of groups, each potentially involved in the development and dissemination of knowledge and norms about climate-related inequalities and appropriate responses. The activities of such groups, accordingly, span a diverse

---

Marx's discussion of that between capitalists and between capital and labor (see Harvey 1981). Late twentieth- and early twenty-first-century associations with the term, while enfolding these earlier connotations distinguishing "private" life from control by "the state," have clearly moved significantly; for example, Kaldor et al. 2012. See also Gidwani 2009.

[62] Gramsci 1971, 243.

[63] Bond (2005) reads the differing emphases in African civil society in terms of these two opposing historical narratives, to suggest the peril within the promise of alignment through the Social Forum process.

[64] Polanyi 2001; Fraser 2011.

[65] Cf. Kaldor et al. 2012.

collection of venues, from the securitized settings of formalized participation in official decision making, to public forums for broad-based deliberation or activism, as well as less visible networking and strategy meetings tied to one or both of those types. As Chap. 4 explains, many organizers identify the open settings of the second type, like Klimaforum and the street march described in the Introduction, as crucial ones for the necessary communicative work of enlarging and aligning coalitions for climate justice. In effect, however, that task of connecting climate change with existing concerns, principles, and constituencies is a basic facet of climate justice-oriented organizations' engagements across these differing settings, and others.

That this "articulatory" work[66] is a particularly apparent feature of advocacy and activism for climate justice across their broad contextual and ideological spectra reflects the paucity of dedicated legal, policy, and political forums and tools, as well as the recognition of urgency. Thus, for instance, for different actors and in different milieus, the uneven social geographies of climate change can be understood as issues of human or "nature's" rights, international ecological debt, racial injustice, and more. Such differing (albeit at times overlapping) ways of understanding climate injustice each figure in advocates' efforts to mobilize resources, publics, and institutions through the construction of politically potent narrative frames: defining the problem and embedding it within "causal stories" through which to raise its salience and impel suitable action.[67] As one member of a transnational network pithily characterized her work at the UNFCCC, "our role is to change the frame."[68]

Given the lack of forums for broad-based politics focused on climate change, however, that essential analytical work must also be paired meaningfully with spatial and social practices enabling affected groups and wider publics to participate in defining impacts and their causes, and in constructing responses.[69] Multiple forms of knowledge, as well as strategies of dialogue and presence are both crucial modalities, therefore, in a thoroughgoing politics of connection—a "praxis" combining analysis and action[70]—fit to contest the relations of climatic injustice.

---

[66] Hart 2006, 2013; Featherstone 2011; see Chap. 6.

[67] Layzer 2015.

[68] Interview, Washington, DC, April 2010.

[69] As they have been repeatedly. See, for example, Routledge (2017) for consideration of innovative spatial practices in climate justice mobilization during the Paris COP in 2015.

[70] Loftus 2013.

Conclusion

The connective analyses and practices of climate justice target recognition of the socio-spatial mismatches between responsibility for and impact from climate change, as well as new capacities to address those disproportionalities in a more inclusive, thoroughgoing manner. In orienting themselves relationally, they resonate with perspectives like Massey's. Her propositions for relationally re-imagining space as the fundamental ground of the political are suggestive of the social and to some degree the material grounds for climate justice projects. She advocates[71]:

> that we recognize space as the product of interrelations; as constituted through interactions, from the immensity of the global to the intimately tiny....
>
> [A]s the sphere of the possibility of the existence of multiplicity in the sense of contemporaneous plurality; in which distinct trajectories coexist....
>
> [A]s always under construction ... a product of relations-between, relations which are necessarily embedded material practices which have to be carried out, it is always in the process of being made ... a simultaneity of stories-so-far.

Massey herself seldom deployed those propositions around explicitly *ecological* concerns. Nevertheless, the conditions of climate injustice and related countermovement struggles clearly suggest their socio-ecological roots and extensions, with critical implications about responsibility for harm and possibilities for deliberation, resistance, and alternatives. These can be provisionally sketched in annotations to her propositions.

First, for instance, relational analyses of climate injustice suggest "that we recognize space [*from nation-states to neighborhoods*] as a product of [*socio-ecological*] interrelations; as constituted through [*material as well as social*] interactions, from the immensity of the global to the intimately tiny." Clearly, that is, these analyses must reflect the interrelations and interactions of a fossil-fueled economy and its industrial ecology, embedded in ongoing histories of ecological and social exploitation, co-producing uneven development and life chances along with climate change. Second, "the existence of multiplicity in the sense of contemporaneous plurality; in which distinct trajectories [*of differentially affected groups and their avowed and potential*

---

[71] Massey 2005a, 9.

allies] coexist" suggests the presence of many formed and potential constituencies of a politics of climate justice, and the possibility of novel and expanding alliances among them. Third, that space might be "always under construction … a product of relations-between" underscores the crucial importance of those alliances, as well as the official and informal power dynamics with which they must contend. And finally, that those same relations "are necessarily embedded in material practices which have to be carried out," and "always in the process of being made" recalls the more-than-human constitutivities of social order, and the ability of human collectives to quite literally make another world, that could better reflect lessons and visions gleaned from many distinct "stories-so-far."

As Massey observed, however, with respect to the interrelations, multiplicities, and material practices she examined, such possible connections are not inevitably made. The following chapter chronicles two attempts to substantiate some of the constitutive connections of climate injustice as legally recognizable ties of responsibility, and the ways in which divergent interpretations have limited their effects.

Bibliography

Agnew, J. (1994). The Territorial Trap: The Geographical Assumptions of International Relations Theory. *Review of International Political Economy, 1*(1), 53–80.

Arctic Council, Assessment, A. C. I., Monitoring, A., Committee, I. A. S., & others. (2005). *Arctic Climate Impact Assessment*. New York: Cambridge University Press.

Bergmann, L. (2011). *Presentation to the Department of Geography of the University of Washington*. Seattle: University of Washington.

Bergmann, L. (2013). Bound by Chains of Carbon: Ecological–Economic Geographies of Globalization. *Annals of the Association of American Geographers, 103*(6), 1348–1370.

Blaikie, P., & Brookfield, H. (1987). *Land Degradation and Society*. London: Routledge Kegan & Paul.

Blaut, J. M. (1993). *The Colonizer's Model of the World: Geographical Diffusionism and Eurocentric History*. New York/London: Guilford Press.

Blomley, N. (1994). *Law, Space, and the Geographies of Power*. New York: Guilford.

Blomley, N., Delaney, D., & Ford, R. (Eds.). (2001). *The Legal Geographies Reader: Law, Power and Space*. Hoboken: Wiley.

Bond, P. (2005). Gramsci, Polanyi and Impressions from Africa on the Social Forum Phenomenon. *International Journal of Urban and Regional Research, 29*(2), 433–440.

Bond, P. (2011). *The Politics of Climate Justice*. London: Verso and Pietermaritzburg: University of KwaZulu-Natal Press.

Bullard, R. D. (2008). *Dumping in Dixie: Race, Class, and Environmental Quality* (3rd ed.). Boulder: Westview Press.

Chakrabarty, D. (2009). The Climate of History: Four Theses. *Critical Inquiry, 35*(2), 197–222. https://doi.org/10.1086/596640.

Cheung, William, W. L., Vicky W. Y. Lam, Jorge L. Sarmiento, Kelly Kearney, R. E. G., Watson, Dirk Zeller, & Daniel Pauly. (2010). Large-Scale Redistribution of Maximum Fisheries Catch Potential in the Global Ocean Under Climate Change. *Global Change Biology, 16*(1), 24–35.

Cheung, William, W. L., John Dunne, Jorge L. Sarmiento, & Daniel Pauly. (2011). Integrating Ecophysiology and Plankton Dynamics into Projected Maximum Fisheries Catch Potential under Climate Change in the Northeast Atlantic. *ICES Journal of Marine Science, 68*(6), 1008–1018.

Cheung, William, W. L., Jorge L. Sarmiento, John Dunne, Thomas L. Frölicher, Vicky, W. Y. Lam, M. L., Deng Palomares, Reg Watson, & Daniel Pauly. (2013). Shrinking of Fishes Exacerbates Impacts of Global Ocean Changes on Marine Ecosystems. *Nature Climate Change, 3*(3), 254–258.

Climate Justice Now! (2010). *CJN! Network Members (as at November 2010)*. Available at: https://web.archive.org/web/20150901040407/http://www.climate-justice-now.org:80/category/climate-justice-movement/cjn-members/

CorpWatch. (2000). *Alternative Summit Opens with Call for Climate Justice*. Retrieved March 14, 2015, from http://www.corpwatch.org/article.php?id=333

Costa, L., Rybski, D., & Kropp, J. P. (2011). A Human Development Framework for CO_2 Reductions. *PLoS One, 6*(12), e29262. https://doi.org/10.1371/journal.pone.0029262.

Crutzen, P. J., & Stoermer, E. F. (2000). Global Change Newsletter. *The Anthropocene, 41*, 17–18.

Dalby, S. (2002). *Environmental Security*. Minneapolis: University of Minnesota Press.

Dalby, S. (2009). *Security and Environmental Change*. Cambridge: Polity.

Delaney, D. (2003). *Law and Nature*. London: Cambridge University Press.

Delaney, D. (2010). *The Spatial, the Legal and the Pragmatics of World-Making: Nomospheric Investigations*. Abingdon: Routledge.

Derman, B. B. (2014). Climate Governance, Justice, and Transnational Civil Society. *Climate Policy, 14*(1), 23–41.

Dryzek, J. S., Downes, D., Hunold, C., Schlosberg, D., & Hernes, H.-K. (2003). *Green States and Social Movements: Environmentalism in the United States, United Kingdom, Germany, and Norway: Environmentalism in the United States, United Kingdom, Germany, and Norway*. Oxford: Oxford University Press.

Edwards, M. (2009). *Civil Society*. Cambridge/Malden: Polity Press.

Escobar, A. (1995). *Encountering Development: The Making and Unmaking of the Third World*. Princeton: Princeton University Press.

Featherstone, D. (2011). On Assemblage and Articulation. *Area, 43*(2), 139–142.

Fraser, N. (2011). Marketization, Social Protection, Emancipation: Toward a Neo-Polanyian Conception of Capitalist Crisis. In C. Calhoun & G. Derluiguian (Eds.), *Business as Usual: The Roots of the Global Financial Meltdown* (pp. 137–158). New York: New York University Press.

Gardiner, S. M. (2006). A Perfect Moral Storm: Climate Change, Intergenerational Ethics and the Problem of Moral Corruption. *Environmental Values, 15*(3), 397–413.

Gardiner, S. (2011). Climate Justice. In J. S. Dryzek, R. B. Norgaard, & D. Schlosberg (Eds.), *The Oxford Handbook of Climate Change and Society* (pp. 309–322). Oxford: Oxford University Press.

Gidwani, V. (2009). Civil Society. In D. Gregory, R. Johnston, & G. Pratt (Eds.), *The Dictionary of Human Geography*. Wiley-Blackwell.

Gill, S. (1995). Globalisation, Market Civilisation, and Disciplinary Neoliberalism. *Millenium: Journal of International Studies, 24*(3), 399.

Gloppen, S., & St. Clair, A. (2012). Climate Change Lawfare. *Social Research: An International Quarterly, 79*(4), 899–930.

Gramsci, A. (1971). *Selections from the Prison Notebooks* (Q Hoare & G. N. Smith, Eds.). New York: International Publishers Co.

Haines, A., & Ebi, K. (2019). The Imperative for Climate Action to Protect Health. *New England Journal of Medicine, 380*(3), 263–273. https://doi.org/10.1056/NEJMra1807873.

Haraway, D. (2015). Anthropocene, Capitalocene, Plantationocene, Chthulucene: Making Kin. *Environmental Humanities, 6*(1), 159–165.

Haraway, D., Ishikawa, N., Gilbert, S. F., Olwig, K., Tsing, A. L., & Bubandt, N. (2016). Anthropologists Are Talking—About the Anthropocene. *Ethnos, 81*(3), 535–564. https://doi.org/10.1080/00141844.2015.1105838.

Hart, G. (2006). Denaturalizing Dispossession: Critical Ethnography in the Age of Resurgent Imperialism. *Antipode, 38*(5), 977–1004.

Hart, G. (2013). Gramsci, Geography, and the Languages of Populism. In E. Michael, H. Gillian, K. Stefan, & L. Alex (Eds.), *Gramsci: Space, Nature, Politics* (pp. 301–320). West Sussex: Wiley-Blackwell.

Harvey, D. (1981). The Spatial Fix–Hegel, von Thunen, and Marx. *Antipode, 13*(3), 1–12.

Harvey, D. (1982). *The Limits to Capital*. Oxford: Blackwell.

Harvey, D. (1996). *Justice, Nature and the Geography of Difference*. Hoboken: Wiley-Blackwell.

Harvey, D. (2007). *A Brief History of Neoliberalism*. Oxford: Oxford University Press.

Harvey, D. (2019). *The Significance of China in the World Economy*. (Video recorded lecture) Democracy at Work. Available at: https://www.democracyatwork.info/acc_the_significance_of_china_in_the_global_economy

Henson, R. (2014). *The Thinking Person's Guide to Climate Change*. Boston: American Meteorological Society.

Hewitt, K. (1983). *Interpretations of Calamity from the Viewpoint of Human Ecology*. Boston: Allen & Unwin.

IPCC (Intergovernmental Panel on Climate Change). (2013, 2014). *Fifth Assessment Report*. Retrieved from http://www.ipcc.ch/report/ar5/

Jafry. Tahseen. (Ed.). (2018). *Routledge Handbook of Climate Justice*. Abingdon: Routledge. https://www.amazon.com/Routledge-Handbook-Climate-International-Handbooks/dp/1138689351

Kaldor, M., Anheier, H., & Glasius, M. (2003). *Global Civil Society*. Cambridge University Press. Retrieved from http://journals.cambridge.org/production/action/cjoGetFulltext?fulltextid=1841140

Kaldor, M., Moore, H. L., Selchow, S., & Murray-Leach, T. (2012). *Global Civil Society 2012: Ten Years of Critical Reflection*. Basingstoke: Palgrave Macmillan.

Keck, M. E., & Sikkink, K. (1998). *Activists Beyond Borders: Advocacy Networks in International Politics*. Ithaca: Cornell University Press.

Keck, M. E., & Sikkink, K. (1999). Transnational Advocacy Networks in International and Regional Politics. *International Social Science Journal, 51*(159), 89–101.

Kolbert, E. (2014). *The Sixth Extinction: An Unnatural History*. New York: Henry Holt and Company.

Kriesi, H. (1995). The Political Opportunity Structure of New Social Movements: Its Impact on Their Mobilization. In C. J. Jenkins & B. Klandermans (Eds.), *The Politics of Social Protest: Comparative Perspectives on States and Social Movements* (pp. 167–198). Minneapolis: University of Minnesota Press.

Latour, B. (1993). *We Have Never Been Modern*. Cambridge, MA: Harvard University Press.

Lawson, V. (2007). Geographies of Care and Responsibility. *Annals of the Association of American Geographers, 97*(1), 1–11.

Layzer, J. A. (2015). *The Environmental Case*. Thousand Oaks: SAGE.

Lazarus, R. J. (2004). *The Making of Environmental Law*. Chicago: University of Chicago Press.

Loftus, A. (2013). Gramsci, Nature, and the Philosophy of Praxis. In E. Michael, H. Gillian, K. Stefan, & L. Alex (Eds.), *Gramsci: Space, Nature, Politics* (pp. 178–196). West Sussex: Wiley-Blackwell.

Lohmann, L. (2008). Carbon Trading, Climate Justice and the Production of Ignorance: Ten Examples. *Development, 51*(3), 359–365. https://doi.org/10.1057/dev.2008.27.

Mahoney, C. (2008). *Brussels Versus the Beltway: Advocacy in the United States and the European Union*. Washington, DC: Georgetown University Press.

Massey, D. B. (2005a). *For Space*. London: SAGE.

Massey, D. (2005b). London Inside-Out. *Soundings-London-Lawrence and Wishart, 32*, 62.

McCann, M. W. (1994). *Rights at Work: Pay Equity Reform and the Politics of Legal Mobilization*. Chicago: University of Chicago Press.

Meyer, D. S., & Minkoff, D. C. (2004). Conceptualizing Political Opportunity. *Social Forces, 82*(4), 1457–1492.

Miller, B. (1994). Political Empowerment, Local–Central State Relations, and Geographically Shifting Political Opportunity Structures: Strategies of the Cambridge, Massachusetts, Peace Movement. *Political Geography, 13*(5), 393–406.

Mitchell, T. (2011). *Carbon Democracy: Political Power in the Age of Oil*. London: Verso Books.

Moore, J. W. (2015). *Capitalism in the Web of Life: Ecology and the Accumulation of Capital*. London: Verso Books.

NAACP (National Association for the Advancement of Colored People). (2012). *Coal Blooded Action Toolkit*. Available at: http://action.naacp.org/page/-/Climate/Coal_Blooded_Action_Toolkit_FINAL_FINAL.pdf

Newell, P. (2005). Climate for Change? Civil Society and the Politics of Global Warming. In F. Holland et al. (Eds.), *Global Civil Society Yearbook* (pp. 90–120). London: SAGE.

Nixon, R. (2011). *Slow Violence and the Environmentalism of the Poor*. Cambridge, MA: Harvard University Press.

Osofsky, H. M. (2006). The Inuit Petition as a Bridge? Beyond Dialectics of Climate Change and Indigenous Peoples' Rights. *American Indian Law Review, 31*(2), 675–697.

Oxfam International. (2015). *Extreme Carbon Inequality*. Oxfam International. Retrieved June 11, 2019, from https://www.oxfam.org/en/research/extreme-carbon-inequality

Parenti, C. (2012). *Tropic of Chaos: Climate Change and the New Geography of Violence*. New York: Nation Books.

Patz, J. A., Gibbs, H. K., Foley, J. A., Rogers, J. V., & Smith, K. R. (2007). Climate Change and Global Health: Quantifying a Growing Ethical Crisis. *EcoHealth, 4*(4), 397–405.

Peet, R. (2003). *Unholy Trinity: The IMF, World Bank and WTO*. London/New York: Zed Books.

Pettit, J. (2004). Climate Justice: A New Social Movement for Atmospheric Rights. *IDS Bulletin, 35*(3), 102–106. https://doi.org/10.1111/j.1759-5436.2004.tb00142.x.

Polanyi, K. (2001). *The Great Transformation: The Political and Economic Origins of Our Time* (2nd ed.). Boston: Beacon Press.

Renton, A. (2011, January 2). *India's Hidden Climate Change Catastrophe*. Retrieved March 16, 2015, from http://www.independent.co.uk/environment/climate-change/indias-hidden-climate-change-catastrophe-2173995.html

Riahi, K., Rao, S., Krey, V., Cho, C., Chirkov, V., Fischer, G., … Rafaj, P. (2011). RCP 8.5—A Scenario of Comparatively High Greenhouse Gas Emissions. *Climatic Change, 109*(1), 33.

Robbins, P. (2004). *Political Ecology: A Critical Introduction*. Malden: Blackwell Pub.

Robine, J.-M., Cheung, S. L. K., Le Roy, S., Van Oyen, H., Griffiths, C., Michel, J.-P., & Herrmann, F. R. (2008). Death Toll Exceeded 70,000 in Europe During the Summer of 2003. *Comptes Rendus Biologies, 331*(2), 171–178. https://doi.org/10.1016/j.crvi.2007.12.001.

Routledge, P. (2011). Translocal Climate Justice Solidarities. In J. S. Dryzek, R. B. Norgaard, & D. Schlosberg (Eds.), *The Oxford Handbook of Climate Change and Society* (pp. 384–398). Oxford: Oxford University Press.

Routledge, P. (2017). *Space Invaders: Radical Geographies of Protest*. London: Pluto Press.

Rutherford, S. (2007). Green Governmentality: Insights and Opportunities in the Study of Nature's Rule. *Progress in Human Geography, 31*(3), 291–308.

Salamon, L. M., & Sokolowski, S. W. (2004). *Global Civil Society: Dimensions of the Nonprofit Sector*. Baltimore: Johns Hopkins Center for Civil Society Studies.

Salamon, L. M., Sokolowski, S. W., & List, R. (2003). *Global Civil Society: An Overview*. Baltimore: Johns Hopkins Center for Civil Society Studies.

Schlosberg, D. (2007). *Defining Environmental Justice: Theories, Movements, and Nature*. Retrieved from http://philpapers.org/rec/SCHDEJ

Shearer, C. (2011). *Kivalina: A Climate Change Story*. Chicago: Haymarket Books.

Smith, N. (2003). *American Empire: Roosevelt's Geographer and the Prelude to Globalization*. Berkeley: University of California Press.

Smith, N. (2008). *Uneven Development: Nature, Capital, and the Production of Space*. Athens: University of Georgia Press.

Sparke, M. (2005). *In the Space of Theory: Postfoundational Geographies of the Nation-State*. Minneapolis: University of Minnesota Press.

Sparke, M. B. (2006). A Neoliberal Nexus: Economy, Security and the Biopolitics of Citizenship on the Border. *Political Geography, 25*(2), 151–180.

Sparke, M. (2012). *Introducing Globalization: Ties, Tensions, and Uneven Integration*. Hoboken: John Wiley & Sons.

Swyngedouw, E. (2007). Impossible "Sustainability" and the Postpolitical Condition. In R. Krueger & D. Gibbs (Eds.), *The Sustainable Development Paradox: Urban Political Economy in the United States and Europe* (pp. 13–40). New York: Guilford Press.

Swyngedouw, E. (2010). Apocalypse Forever? Post-Political Populism and the Spectre of Climate Change. *Theory, Culture & Society, 27*(2–3), 213–232.

Terry, G. (2009). No Climate Justice Without Gender Justice: An Overview of the Issues. *Gender and Development, 17*(1), 5–18.

Turk, A. T. (1976). Law as a Weapon in Social Conflict. *Social Problems, 23*(3), 276–291.

U.S. Global Change Research Program. (2018). *Fourth National Climate Assessment, Volume II: Impacts, Risks, and Adaptation in the United States.* https://nca2018.globalchange.gov/

UNDP (United Nations Development Program). (2009). *Human Development Report 2009: Overcoming Barriers: Human Mobility and Development.* New York: United Nations Development Program.

UNEP (United Nations Environment Program). (2014). *Environmental Data Explorer.* Retrieved March 15, 2015, from http://geodata.grid.unep.ch/extras/posters.php

Van Der Heijden, H.-A. (2006). Globalization, Environmental Movements, and International Political Opportunity Structures. *Organization & Environment, 19*(1), 28–45.

Wallace, R. (2014, January). *Whipsaw of Damocles: Are Climate Change and Pandemic Influenza Related?* Presented at the Simpson Center for the Humanities, University of Washington.

WBGU (German Advisory Council on Global Change). (2009). *Solving the Climate Dilemma: The Budget Approach.* Berlin: German Advisory Council on Global Change.

WHO (World Health Organization). (2007). *Climate Change: Quantifying the Health Impact at National and Local Levels.* Geneva: World Health Organization. Available at: https://apps.who.int/iris/bitstream/handle/10665/43708/9789241595674_eng.pdf?sequence=1

WHO (World Health Organization). (2014). *Quantitative Risk Assessment of the Effects of Climate Change on Selected Causes of Death, 2030s and 2050s.* Geneva: World Health Organization.

WHO (World Health Organization). (2019). *Climate Change.* WHO. Available from: http://www.who.int/heli/risks/climate/climatechange/en/

Wildcat, D. R. (2013). Introduction: Climate Change and Indigenous Peoples of the USA. *Climatic Change, 120*(3), 509–515. https://doi.org/10.1007/s10584-013-0849-6.

WRI (World Resources Institute). (2013). *New Report Connects 2012 Extreme Weather Events to Human-Caused Climate Change.* Retrieved February 20, 2015, from http://www.wri.org/blog/2013/09/new-report-connects-2012-extreme-weather-events-human-caused-climate-change

Zemans, F. K. (1983). Legal Mobilization: The Neglected Role of the Law in the Political System. *The American Political Science Review, 77*(3), 690–703.

Part I
International Laws and Institutions

Climate Wrongs and Human Rights

2

As climate-related harms became more widespread and more clearly disproportionate, and governmental responses continued to lag, affected groups and their allies began to explore possibilities for remedy through legal action. Legal mobilization, after all, figures among the more widely recognized and well-theorized tools by which minorities and marginalized groups seek protection or redress in the face of social exclusion, violence, oppression, and negative market "externalities."[1] Legal mobilization in response to disproportionate climate-related harm has emerged in tandem, though not always in connection, with wider social and political movements and institutional advocacy for "climate justice."[2] Claimants and theorists of legal mobilization for climate protection and redress have adopted a wide range of approaches across many venues.[3] Perhaps because of the broad salience they maintain in governance, policy, and political movements, however, human rights theories have been particularly prominent in legal advocacy as well as in wider activism and scholarship.[4]

---

[1] See, for example, Zemans 1983; McCann 1994.

[2] For example, Bond 2011; Shearer 2011; Derman 2013; Mary Robinson Foundation—Climate Justice 2017.

[3] Abate 2010; Averill 2010; Gloppen and St. Clair 2012; Werksman 2011.

[4] For example, Sachs 2008; ActionAid et al. 2010a, b; Humphreys 2010; McInerney-Lankford et al. 2011; Derman 2014; Atapattu 2015; Mary Robinson Foundation—Climate Justice 2017.

© The Author(s) 2020
B. B. Derman, *Struggles for Climate Justice*,
https://doi.org/10.1007/978-3-030-27965-3_2

Thus, as Rajamani argued, it was *"axiomatic,"* by 2010, "that the climate impacts documented by the Intergovernmental Panel on Climate Change are likely to undermine the realisation of a range of protected human rights."[5] Since then, as climate impacts have proliferated and gained force, climate rights talk has as well. Its material significance for affected groups remains doubtful, however, since legal analyses have yet to compel or inspire international remedy, protection, or redress for climate-related environmental harm.[6] Arguably, rather, official responses to claims of strong responsibility for direct and climate-related environmental harms have undercut the potential for legally mandated compensatory action.

This chapter exposes some of the central conditions and implications of this situation, using two foundational legal analyses of climate change and human rights. Both tested the question of whether or not causing climate change should be understood as actually *violating* rights, an interpretation which could imply legal responsibility and ultimately compel corresponding redress by identifiable actors. The analyses are the Inuit Circumpolar Council (ICC) petition to the Inter-American Commission on Human Rights (IACHR), together with the Commission's response, and the Maldives submission to a study on human rights and climate change conducted by the United Nations (UN) Human Rights Council (HRC), together with a report synthesizing that study, issued by the Office of the High Commissioner for Human Rights (OHCHR).[7]

The following sections analyze the construction of and response to the ICC's and Maldives' claims, in the context of their specific fora and in relation to evolving understandings of responsibility for climate-related harm in scientific, political, and public discourse. The claims demonstrate that when scientific and traditional knowledges are brought to bear on the interpretation of recognized rights, climate change *can* be seen as violating those rights and state actors understood as culpable. The responses show, however, that the complexity of climate-related harm, which involves multiple human actors, environmental processes, and global ties, has also been understood in legal terms as obscuring responsibility, making claims of rights violations appear insubstantial. The gap between these interpretations highlights the power of foundational analytical and spatial categories ill-fit to novel problems, raising doubt about law's relevance to the inequities of climate change and other harms that mix people, things, and places.

---

[5] Rajamani 2010, 391, emphasis added.

[6] A human rights case in the Netherlands has recently set a domestic precedent (*Urgenda Foundation v. The State of the Netherlands* 2015). It is unclear what implications it may hold for international claims, which are the subject of this analysis.

[7] ICC 2005; OAS 2007; Government of the Maldives 2008; OHCHR 2009.

The OHCHR, in particular, has strengthened its advocacy for rights-based protections from climate harm,[8] but it has not revisited important limiting elements of the analysis it offered in 2009. Reasons exist for it to do so, but issues of venue and geopolitical power would still then remain.

As "cases"—in the social scientific rather than the legal sense—the ICC's and Maldives' efforts and their outcomes represent significant moments in the emerging legal understanding of climate change and the longer process defining human rights. They are among the earliest international climate rights claims to entail extensive claimant argumentation as well as official responses. Those claims and responses have had broad influence, in part because of their settings in international fora.[9] The cases also exemplify legal dimensions of the wider institutional relationalities crucial to the pursuit of climate justice and, potentially, other border-crossing, socio-ecological challenges. These dimensions include issues of causality that distinguish legal reason from other ways of knowing; law's dependence on the spatialized construction and expression of social power; and the utility of legal mobilization for groups otherwise marginalized within existing governance, as knowledge and conditions change.

Complexity, Power, and Counter-Hegemony

In the official responses to the ICC's and Maldives' claims, and others testing legal principles for climate-related harm, the term "complexity" occupies a position of particular importance.[10] It is the "complexity" of these harms, though they may be acknowledged as bearing upon rights, that can prevent recognition of legally responsible human actors. Although climate impacts are indeed complex (in the definitional sense that they can result from multiple identifiable and interacting elements), tracing causality and assigning responsibility for those impacts appears comparatively unproblematic and increasingly tractable in scientific, governance, and journalistic analyses. Official responses to the Maldives and ICC hewed instead to conceptions of identifiable, harmful actions as individual rather than concerted, and unmediated by the participation of non-human elements.

---

[8] E.g. OHCHR 2015.
[9] See, e.g., Burger and Wentz 2015.
[10] Cf. Averill 2010; Gloppen and St. Clair 2012.

Research in legal geographies has identified a pattern of related contradictions between legal and scientific epistemologies, as these different perspectives characterize phenomena.[11] In many instances, non-human entities, combinations of the human and the non-human, and human biology alone readily escape law's categories, often rendering them dubious. When complexity is invoked as a limit to legibility—and by extension responsibility, as in the 2009 OHCHR Report—it marks such points of friction between legal and other ways of understanding the socioecological processes by which anthropogenic climate change arises and concentrates impact in human communities.

As mixtures of human and non-human action, and as novelties in relation to rule and precedent, then, climate-related harms present puzzles for legal reason. At one level, this interpretive open-endedness is familiar in that debates over climate change and human rights recapitulate aspects of those pitting formalist against realist legal philosophies.[12] At another level, though, it bears reiterating that legal interpretation is also intimately tied to the construction, extension, and exercise of power.[13]

Relatedly, scholarship in political ecology suggests the need for a particularly wary appraisal of interpretation in climate justice cases, since it recognizes that power often operates through the social production and construction of the "natural."[14] These processes, moreover, may go unrecognized: what we call "nature" is both discursively constructed and materially produced through the given categories of dominant forms of knowledge and the institutionalized practices they underpin. The reproduction of social inequality and risk connected with environmental conditions is thereby frequently naturalized, and misrecognized as "natural."[15]

Geographic perspectives also highlight the foundational importance of spatial control in defining and facilitating legal order.[16] Just as law requires space in which to be "performed," it also produces specific kinds of spaces, with corollaries in entitlement, exclusion, and other status categories for human subjects. Territories associated with many administrative scales are primary examples of such legal spaces.[17] Territorially delimited, state-contractual conceptions of human rights

---

[11] For example, Delaney 2003; Blomley 2008; Herbert et al. 2013.
[12] Cf. Bix 2005; Mertz et al. 2016.
[13] Cover 1983; Galanter 1974.
[14] Swyngedouw 1999; Robbins 2007, 2012.
[15] Castree 2005.
[16] Blomley 1994; Blomley et al. 2001; Braverman et al. 2014.
[17] Herbert 1997; Delaney 2010.

protections present instances of such consequential spatio-legal categories, potentially insulating emitters from legal consequences associated with climate-related harm in foreign places.

On the other hand, legal mobilization offers counter-hegemonic possibility—albeit conditionally—enabling subaltern claimants to access official power as legal subjects, and channel it toward socially protective, even transformative, ends.[18] Arguably, recent political trends have impelled an increase in legal mobilization around a wide variety of causes and across political spectra. Climate justice legal mobilization can be seen as one among several such forms of "lawfare," which participate in a wider judicialization of politics that has, in certain times and places, transformed legal and social relations.[19] The ICC's and Maldives' claims and their legacy illustrate some of the conditions that have limited the potential of rights-based climate justice legal mobilization, and the implications of those conditions and outcomes for the role of law in addressing the disproportionalities of climate-related harm.

Arctic and Island Pioneers

The IACHR is a regional forum for the evaluation of human rights claims under the auspices of the Organization of American States (OAS). It is distinct from the OAS' adjudicative body (the Inter-American *Court* of Human Rights). In 2005, the ICC, an international NGO representing Inuit communities in Alaska, Canada, Greenland, and Chukotka (Russia) petitioned the IACHR, charging the US federal government with violations of Inuit human rights due to the US' continued, unregulated, plurality contribution to global greenhouse gas emissions.[20] Although it incorporates formal legal analysis, the petition's main signatory and commentators have described the ICC's effort as a creative blurring of established legal categories, aiming at "initiating dialogue" rather than winning an enforceable judgment, which would supersede the IACHR's mandate.[21] The inclusion of photographs of arctic people and landscapes, figures visualizing scientific data and concepts, and excerpts (accompanied by online videos) of interviews with several named claimants support this characterization of the petition as

---

[18] Turk 1976; Zemans 1983; McCann 1994.

[19] Gloppen and St. Clair 2012.

[20] At the time of the petition, the US led all nation-states in cumulative, per capita, and gross annual emissions measures.

[21] Osofsky 2006; Chapman 2010.

communicative in its intent. The IACHR's initial response, in contrast, was terse. In a letter to ICC legal counsel, it stated that "the information provided [in the petition] does not enable us to determine whether the alleged facts would tend to characterize a violation of rights."[22] Subsequently, in 2007, the Commission hosted a hearing in which ICC representatives responded to questions raised by IACHR commissioners, touching on both climate science and human rights.[23]

The ICC Petition analyzes climate impacts in the Arctic in relation to multiple rights recognized in the OAS' American Declaration on the Rights and Duties of Man ("the Declaration" below),[24] including self-determination, the use and enjoyment of traditional lands, personal property, health, life, physical integrity and security, subsistence, residence, movement, and the inviolability of the home. It places climate-related effects on these rights in the context of prior decisions of the Inter-American Court as well as the Indigenous and Tribal Peoples Convention of 1989. Inuit traditional knowledge plays a crucial role in the petition, since it helps to demonstrate how the transformative impacts of climate change on the arctic landscape are having devastating effects on Inuit culture—"culture" being rhetorically central in the Declaration.[25]

The Maldives' intervention took shape through a formal process of international study and reporting under the HRC and OHCHR, though its stated goal was international recognition and compliance under the United Nations Framework Convention on Climate Change (UNFCCC).[26] The OHCHR Report was the centerpiece in a campaign by the Maldives, other members of the Small Island Developing States coalition of UN member countries, and NGOs, to influence the climate treaty through their work in the human rights institutions. Emerging out of the low-lying island nations' dire assessment of their future amidst the halting progress of international climate governance, the group sought to compel stronger international cooperation by framing the existential threat of sea level rise in terms of internationally recognized human rights. The Maldives proposed HRC Resolution 7/24, which requested a formal study of rights and climate change. The proposal passed unanimously, and several member countries and observers submitted briefs as part of the ensuing study, which culminated in 2009. Countries' submissions were then jointly summarized in the OHCHR Report. The Maldives' own extensive submis-

---

[22] Dulitzky 2006, 1.

[23] OAS 2007.

[24] Ninth International Conference of American States 1948.

[25] Culture is prominently placed, for example, in the preamble of the Declaration and as one of the enumerated rights. OAS 2007.

[26] Limon 2009.

sion argued that developed nations bore responsibility for violations of its citizens' rights associated with climate change. The submission utilizes findings from the IPCC and information on the physical geography, economy, and culture of the Maldives to link the emissions of industrial nations with a list of affected rights spanning several international human rights conventions.

Because the HRC study process documented analyses of human rights and climate change, as submitted by several states, and the OHCHR Report summarizes those analyses, the Maldives case exposes differences between countries' views as well as the manner in which those differences were ultimately mediated. Whereas the IACHR commissioners looked to prior decisions and existing official analyses of the substantive meaning of human rights to guide their evaluation of the ICC Petition's claim—finding none to adequately support a claim of violations—the HRC/OHCHR process produced an analysis in the Report which could have served that role in future cases, but likewise eschewed the idea of violation.

If the finding against violation status in the two responses was similar, fundamental commonalities also exist in argumentation in the claims and responses. Despite the differences noted above, the ICC's and Maldives' claims both depend on substantiating two types of connection as ties of legal responsibility: first, between "environmental" processes and human actions (socio-ecological connections), and second, between emitters and affected communities separated by national borders (socio-spatial ones). Articulating those connections as legally legible ties of responsibility constitutes the major argumentative work of the two claims. That work consists, in both cases, of supporting legal reasoning with substantive information from climate science and other non-legal forms of knowledge. These non-legal knowledges play a crucial role: grounding observed and predicted climate change impacts in specific socio-ecological conjunctures of vulnerability in different places and communities, and demonstrating how those situated impacts jeopardize recognized rights.

The logic of the official responses to the two claims also contained fundamental similarities, and these relate directly to the two forms of connection the claims sought to substantiate as implying legal responsibility on the part of high emitters. First, both official responses balked at recognizing the socio-ecological connections of climate change as constituting rights violations, raising the concerns about "complexity" noted above. Second, both responses raised concerns about the international scope of potential human rights impacts, though the OHCHR Report expressed much more circumspection about the claim of violations in relation to this theme than did the IACHR.[27]

---

[27] This is an area in which the OHCHR's analysis has evolved in subsequent years, as discussed below.

The ICC Petition

The argument of the ICC Petition ("the Petition," in what follows) involves three key moves: explicating the impacts of climate change on Inuit people, framing those impacts as affecting human rights, and linking them with actions and omissions by the US government, which contravene obligations based in human rights agreements and other relevant law. Those moves involve incorporating elements of Inuit traditional knowledge (*Inuit Qaujimajatuqangit*, or IQ) and what the ICC calls "western" science with legal sources.

IQ and western science help to substantiate the connections of climate change as ties of legal responsibility. The Petition spends little time on the validity of rights violations that cross national borders, perhaps because many Inuit reside in the US. Ties of responsibility that would encompass Inuit communities elsewhere are inferred from the global extent of climate change (made clear in the scientific studies cited) and the international scope of relevant legal instruments. On the other hand, the Petition explicates in detail the socio-ecological forms of connection by which climate change might be understood as violating rights. It addresses two facets of these connections in particular: first, how environmental impacts associated with climate change are produced socially, through human-caused emissions and their effects on the earth's physical systems, and second, how those impacts are re-socialized via their effects on human populations.

To trace the US' responsibility for harm in Inuit communities across the space and time of climate change, the Petition mobilizes well-supported and disseminated scientific findings, including those published by the US government itself. It explains the anthropogenic intensification of the greenhouse effect, to illustrate that "[g]lobal warming is caused by human activity."[28] Given the persistence of carbon in the atmosphere and the long tail that characterizes its impacts on the earth's climate system, the ICC shows that the great bulk of warming and resulting ecological impacts to date can, in fact, be associated with emissions from the US. Summarizing historical records compiled by the IPCC, the Petition concludes that "U.S. greenhouse gas emissions between 1850 and 2000 are responsible for 0.18 °C (30%) of the observed temperature increase of 0.6 °C during that period,"[29] a share that significantly outstripped any other single nation and the European Union as a whole. Finally, the Inuit are subject to the consequences of US emis-

---

[28] ICC 2005, i, 27.

[29] Ibid, 68–69.

sions by virtue of the global distribution of climate change impact, and particularly so, since, as a raft of scientific studies illustrate, "[g]lobal warming is most severe in the Arctic."[30]

The Petition frames climate change impacts in the Arctic as human rights violations through an accounting of Inuit peoples' deep connection with and reliance upon what have become unstable ecological conditions in the region. Crucially, these strong, and now eroding, socio-ecological ties are both physical and cultural; their integrity is therefore fundamental to the human rights recognized in the Declaration and elsewhere. To substantiate this claim, the Petition first methodically demonstrates the validity and importance of IQ for the continuity of Inuit culture and livelihood, using the testimony of Inuit people, non-Inuit arctic explorers, and health statistics. IQ encodes sophisticated understanding of weather, climate, and the ecological conditions they influence, which in turn supports Inuit practices of hunting, travel, and shelter. These practices are essential for the maintenance of community, health, identity, intergenerational cohesion, and in many cases life itself. It is because of these specifically Inuit imbrications of ecology and society, maintained by IQ, that the Petition argues: "[t]he life and culture of the Inuit are completely dependent on the Arctic environment."[31] The Petition synthesizes the testimony of community elders, who bear special responsibility for the transmission of IQ, together with "western" scientific findings about rapid, dramatic climate change impacts in the Arctic to substantiate its core claim that "[g]lobal warming harms every aspect of Inuit life and culture."[32]

Accordingly, because "the United States is the world's largest contributor to global warming,"[33] and by virtue of US acts and omissions neglecting and obstructing the regulation of greenhouse gases in breach of what the Petition argues are its legal obligations,[34] the ICC concludes: "[t]he effects of global warming constitute violations of Inuit human rights, for which the United States is responsible."[35]

---

[30] Ibid, 33.

[31] Ibid, i, 33.

[32] Ibid, ii, 35.

[33] Ibid, iii, 68.

[34] This interpretation is based on the Petition's analysis of US domestic policy and its role in international climate negotiations.

[35] Ibid, iii, 70.

The Maldives Submission to the Human Rights Council Study

Like the Petition, the Maldives submission to the HRC study ("the Submission," in what follows) combines legal and non-legal knowledge to frame the global and socio-ecological connections of climate-related harm as constituting ties of responsibility for rights violations. Whereas the concept of culture provides a lynchpin in the ICC's claim, statehood plays a similar role for the Maldives. Correspondingly, whereas the Petition drew on Inuit traditional knowledge to cement crucial logical connections, the Submission employs statistical and numerical data on the Maldives' physical and economic geography.

The Maldives gives much more consideration than the ICC to the legal bases of extraterritorial rights protection (mirroring global climatic connections). It draws on multiple sources of law to argue that, in the face of anthropogenic sea level rise (SLR), high-emitting countries must protect Maldivians through mitigation and support for adaptation. SLR stands at the center of the Submission's analyses of socio-ecological impacts on rights: SLR poses an existential threat to the Maldives territory and therefore, it argues, to statehood, with disastrous consequences for the full panoply of its citizens' human rights (see below).

The Submission provides a detailed analysis of the social consequences of local environmental impacts resulting from climate change, using these to map current and future climate change impacts identified by the IPCC onto specific rights and international legal instruments. The IPCC finds, for instance, that rising temperatures from climate change impact fisheries. Catch has diminished in the Maldives; one among other "[c]hanges in traditional fishing livelihood and commercial fishing," which in turn threatens the right to means of subsistence enshrined in the International Covenant on Economic, Social and Cultural Rights (ICESCR).

Ultimately, the Maldives' social and economic welfare and its sovereignty are seen to depend upon the integrity of territory and resources within and surrounding its low-lying atoll islands, which are gradually but inexorably being lost to the rising sea. The Submission therefore argues that the possibility of territorial erasure from SLR threatens every right Maldivians hold under international law[36]:

> In the long-term, unchecked sea-level rise will inundate the whole of the Maldives. The extinction of their State would violate the fundamental right of Maldivians to possess nationality and the right of the Maldives people to self-determination. Without

---

[36] Government of the Maldives 2008, 21.

land or State, the most basic rights to life, liberty, and security of person, to possess property, to work and to leisure, to an adequate standard of living, to participate in the cultural life of the community, cannot be realized. The loss of land and State renders all other rights, political and civil as well as economic, cultural, and social rights, unattainable.

The Maldives supports its claim for developed countries' obligations to protect its citizens' human rights using the International Covenant on Civil and Political Rights (ICCPR), the ICESCR, and customary law. It uses the ICCPR to build an argument for obligations on the part of high-emitting nations by linking its low-lying island geography with a territorial conception of statehood. The Submission cites commentary on and interpretations of the ICCPR's right to self-determination in particular, which obligate signatories to respect and promote the rights of citizens of other states and beyond their borders, because territorial sovereignty figures in achieving self-determination. It argues that these obligations imply states are legally required to mitigate greenhouse gas emission to scientifically agreed safe levels, and that appropriate targets should be mandated under the UNFCCC. Here the Maldives' reading of the right to self-determination provides a means of countering objections to cross-border relations of responsibility for rights: it grounds those relations legally as relations among states, pointing to the international system for redress.

From the ICESCR, the Submission highlights provisions specifying that states "take steps, individually and through international assistance and cooperation, especially economic and technical, to the maximum of [their] available resources, with a view to achieving progressively the full realization of the rights recognized in the present Covenant."[37] It notes that, according to ICESCR Committee interpretation, "[t]he economically developed States parties have a special responsibility and interest to assist the poorer developing States...."[38] The Maldives argues that therefore "[c]limate change, because of its trans-boundary nature and the acute threat it poses to economic, social, and cultural rights among vulnerable populations, is an issue that implicates the responsibility of all State parties to cooperate."[39] It specifies corresponding duties of states and the international community to mitigate greenhouse gas emissions, adhere by the obligations of climate change agreements, and provide aid for adaptation efforts in the Maldives and elsewhere.

---

[37] Ibid, 76.
[38] Ibid, 77.
[39] Ibid, 77.

Citing the principles of non-discrimination and responsibility to protect, the Maldives argues that customary law also demands developed nations bear responsibility for emissions within their borders inasmuch as those emissions impinge on human rights, "regardless of the location of the beneficiaries of those rights."[40] "Customary law," it argues, "emphasizes the protection of human dignity, without limitations based on nationality."[41]

Finally, the Submission argues that the UNFCCC and its Kyoto Protocol offer a "good preliminary framework"[42] for cooperative action through which states might meet their legal obligations for climate protections. It lauds the inclusion of the principle of Common but Differentiated Responsibilities (CBDR), the precautionary principle, mitigation targets for developed countries, the requirement of technical and adaptation assistance, and the designation of particularly vulnerable Least Developed Countries as deserving of special consideration. On the other hand, the Submission notes that several states have breached their Kyoto commitments, and that spatially distributed mitigation programs like the Protocol's Clean Development Mechanism entail risk to rights.[43]

Like the ICC, the Maldives constructed a legal argument for interpreting climate change impacts as rights violated by identifiable actors that is based in its own circumstances, but entails broader implications. Official responses to the two claims did not identify clear, commensurate forms or levels of responsibility amidst the complex socio-ecological and extraterritorial connections of climate-related harm.

Official Replies

In 2007 the IACHR heard testimony from Sheila Watt-Cloutier (lead signatory to the ICC Petition) and ICC counsel, expanding on themes from the Petition. A panel of three commissioners then posed questions, to which the presenters responded.[44] In addition to inquiring whether petitioners had exhausted domestic legal remedies prior to approaching the IACHR, and what positive responses to climate change existed on which it might base recommendations to the OAS states, the

---

[40] Ibid, 79.
[41] Ibid, 78.
[42] Ibid, 81.
[43] Ibid, 79–82.
[44] OAS 2007.

commissioners posed two questions that indicated specific concerns that had led them to dismiss the Petition initially. These pertained to the apportionment of responsibility among multiple states and the violation of Inuit human rights through climate change impacts.

Commissioner Paulo Sergio Pinheiro asked "how the commission can attribute responsibility to a whole region or to a state or even to states which are not members of the OAS....[h]ow to divide, how to share, this responsibility."[45] Commissioner Victor Abramovich inquired[46]:

> Is there a precise form in which the impact you have described very well on fundamental rights can be tied to the actions or omissions of the particular states? ... [I]n all cases ... considered by the Inter-American system, there have existed direct actions ... or the failure to act by the state in the face of a concrete situation, for example ... forestry in an Indigenous territory. Now, the problem you are laying out, without doubt, links to state and non-state actors, but the relationship is much ... less direct. So, I would like clarification about how there can be a relationship—not just any relationship, a legal relationship, a relationship of responsibility—of the states for violations of the rights that you have very clearly described.

ICC representatives responded to Pinheiro's question by citing the principle of CBDR found in the UNFCCC and other treaties (see below), and argued that human rights law does not actually require such an apportionment among jointly responsible parties. Addressing Abramovich's concern, they pointed to the ability states already exercise, and should under human rights law, to regulate the actions of private entities within their borders, but did not address the traceability of human action through socio-ecological connections to harmful environmental events. Counsel also noted that no legal remedies are available to individuals who suffer climate change-related harm under US or Canadian law, or via the UNFCCC.[47]

In subsequent years the IACHR has addressed climate change from a human rights perspective at least three times. Those statements have progressed from noting the existence of a relation between the two, to reiteration of language from the OHCHR Report, discussed below, to encouraging OAS members to advocate for robust human rights language in the UNFCCC Paris Accord.[48]

---

[45] Ibid, at 33:24.

[46] OAS 2007. Abramovich spoke in Portuguese, beginning at 34:30. The translated excerpt here comes from Chapman (2010, 38).

[47] OAS 2007.

[48] OAS 2008, 2011, 2015.

Similar circumspection about the attribution of responsibility proved consequential in the OHCHR Report, albeit amidst a clear acknowledgment and substantive discussion of the implications climate change holds for human rights. Indeed,
several themes developed throughout the Report clearly suggest the influence of
the Maldives and other countries who initiated the study. Most notably, the Report
states that climate change poses "implications for the enjoyment of" a wide range
of human rights, and that human rights law places duties on states to respond, both
domestically and through "international cooperation."[49] The Report discusses implications of climate impacts for rights to life, food, water, adequate housing, and
self-determination. It also notes differential impacts in relation to the rights of
women, children, and Indigenous peoples, and warns that rights may be affected by
climate change-induced displacement, conflict, and compromised security, as well
as governmental responses.

The Report attributes states' domestic obligations in connection with climate
change to their recognized duties to pursue progressive realization of economic,
social, and cultural rights, and promote access to information and participation in
decision making, for their citizens. It also calls on states to incorporate rights as
"guiding principles for [climate change] policy-making."[50] In enumerating these
duties and the "implications" for rights described above, the Report provided a new
level of institutional affirmation for understanding and responding to climate
change impacts in terms of human rights.

At the same time, major facets of the Maldives argument were elided, softened,
or rejected in the OHCHR Report. For one, the Report lacks an analysis of international obligations commensurate with its discussion of domestic duties. Rather, it
defers specification of international duties to the UNFCCC, arguing that "international human rights law *complements* the United Nations Framework Convention
on Climate Change"[51] rather than, for instance, compelling the latter to adopt a
rights-oriented approach.

The Report does ground an obligation of states to international cooperation in
multiple sources of law and follows the Maldives in noting ICESCR commentary
which specifies that developed countries have "particular responsibility and interest to assist the poorer developing States."[52] It also states that, "[w]hile there is no
clear precedence to follow, it is clear that insofar as climate change poses a threat

---

[49] OHCHR 2009, multiple locations.

[50] Ibid, 24–27.

[51] Ibid, 26, emphasis added.

[52] Ibid, 24.

to the right of peoples to self-determination, States have a duty to take positive action, individually and jointly, to address and avert this threat."[53] It does not, however, return to that theme when enumerating or summarizing states' extraterritorial obligations. Nor, relatedly, did the Report link the protection of self-determination or other rights in any way with a duty to mitigate greenhouse gas emissions, which would constitute the only means of actually arresting, at some future date, deleterious climate impacts such as rising seas.

Further, and more expansively than the IACHR commissioners, the OHCHR Report expresses significant doubt about assigning responsibility amidst the complexities of anthropogenic climate change and climate-related harm. Those doubts, together with the prospective character of the most dire climate-related harms, marked an analytical boundary between classifying climate change as posing "implications for" and constituting "violations of" human rights. The Report argues[54]:

> While climate change has obvious implications for the enjoyment of human rights, it is less obvious whether, and to what extent, such effects can be qualified as human rights violations in a strict legal sense. Qualifying the effects of climate change as human rights violations poses a series of difficulties. First, it is virtually impossible to disentangle the complex causal relationships linking historical greenhouse gas emissions of a particular country with a specific climate change-related effect, let alone with the range of direct and indirect implications for human rights. Second, global warming is often one of several contributing factors to climate change-related effects, such as hurricanes, environmental degradation and water stress. Accordingly, it is often impossible to establish the extent to which a concrete climate change-related event with implications for human rights is attributable to global warming. Third, adverse effects of global warming are often projections about future impacts, whereas human rights violations are normally established after the harm has occurred.

The "difficulties" enumerated in this except clearly overlap with those about which IACHR commissioners asked in the ICC's hearing and those raised in other legal discussions.[55]

Subsequently, HRC Resolution 10/4, which summarized the Report's content in advance of the landmark 2009 climate negotiations in Copenhagen, moderated the Maldives' claims still further.[56] In the event, the Maldives and its allies' bid to establish a more ambitious agreement there based in part on human rights obligations

---

[53] Ibid, 13.
[54] Ibid, 20.
[55] See, for example, Averill 2010.
[56] HRC 2009; Cf. Knox 2009.

did not prevail. Rights went unmentioned in the voluntaristic vision of global climate governance that took hold in Copenhagen and has since recast the global climate regime in important ways, which have sidelined both legal obligation and scientific guidance (see Chap. 3).[57]

Divergent Interpretations and Limited Progress

As described above, the 2009 OHCHR Report emerged from a joint drafting process reflecting the participation of HRC member states, many of which contributed their own submissions to the Council's study. The availability of those submissions allows for the tracing of elements within the Report to individual countries' statements. That tracing makes clear that the Maldives' was not the only analysis of climate change in the context of human rights to consequentially shape the Report, and that drafters faced a challenge in synthesizing notably divergent interpretations.

The Report excerpt quoted above on the difficulty of "[q]ualifying the effects of climate change as human rights violations," for instance, closely followed language in the US submission to the study.[58] The US also argued at some length that "the view that an environment-related human right exists …does not have a basis in international law."[59] That view had been prominent in the Maldives' submission.[60] Rejecting environmental rights, of course, logically undercuts arguments for rights-based protection against socio-ecological harms. The Report charts a middle ground with respect to this issue, stating that "[w]hile the universal human rights treaties do not refer to a specific right to a safe and healthy environment, the United Nations human rights treaty bodies all recognize the intrinsic link between the environment and the realization of a range of human rights."[61] Numerous national constitutions do include environmental rights.

The US also argued for an understanding of human rights in general as obliging states to protect their own citizens within their borders. For this and other reasons,

---

[57] Rights were not mentioned in the Copenhagen Accord, and the voluntary commitments it recorded fell far short of a level necessary to save the island nations. Commitments under the Paris Agreement of 2015, similarly, are inadequate to realize the treaty's founding objective: "stabilization of greenhouse gas concentrations in the atmosphere at a level that would prevent dangerous anthropogenic interference with the climate system (United Nations 1992)." UNFCCC 2009; Rogelj et al. 2016; Lesnikowski et al. 2016.

[58] Government of the United States 2008, 4–5; Cf. Knox 2009.

[59] Ibid, 3.

[60] Government of the Maldives 2009, 12–14.

[61] OHCHR 2009, 7.

it expressed skepticism about the appropriateness of adopting a human rights framework to address climate change.[62]

Legal scholar John Knox speculated on the choices of the Report drafters in mediating among such divergent views as those of the Maldives and the US, noting that accusing the most powerful states of violating human rights could "distract from the need to win their consent to an effective climate agreement."[63] The necessity for such calculations underscores the embeddedness of legal interpretation in political processes, which is particularly evident in international law but consequential more generally.[64]

In subsequent years, the OHCHR has engaged more directly with the UNFCCC, and its statements about the human rights implications of climate change have become both stronger and more influential. A few specific indicators of change bear mentioning. First, the preface to the UNFCCC treaty negotiated in 2015 includes language on human rights. This is a landmark of sorts, even given that the invocation is probably much too general to exert significant effects on implementation.[65] Second, recent analyses by the OHCHR and other UN bodies specify that mitigation *is* an obligation of signatories to the major human rights conventions, and one which they hold to rights bearers abroad.[66] Finally, those analyses now reference violations of human rights as documented effects of certain mitigation and adaptation programs and possible outcomes of relocation and resettlement.

These, however, are not the type of climate-related impacts the ICC sought to substantiate as violations, nor those with which the Maldives was most concerned. Both were interested primarily in harm from environmental change and events. Moreover, the socio-ecological processes and outcomes the ICC and the Maldives described as human rights violations are now only more widespread.

Legal claims within states have recently resulted in a few judgments of responsibility on the part of governments to citizens for such climate-related environmental harms, and exploration of potential liability on the part of corporations.[67] The need remains at the international level, however, for means of protection and redress, if not also the assignment of legal responsibility. Without these, the most

---

[62] Government of the United States 2008, 6.

[63] Knox 2009, 489–490.

[64] See, for example, Scheingold 1974; Bell 1980; Merry 2006.

[65] See Mayer 2016.

[66] For example, OHCHR 2015; Burger and Wentz 2015.

[67] Burger and Wentz 2015, 20, 23; *Urgenda Foundation* vs. *The State of the Netherlands*; Petition Requesting for Investigation of the Responsibility of the Carbon Majors for Human Rights Violations or Threats of Violations Resulting from the Impacts of Climate Change.

vulnerable and lowest emitting states continue to bear ad hoc responsibility for direct climate impacts on their citizens, and displaced populations will have to suffer (and have survived) such impacts before they may benefit from mandated protection in resettlement or relocation. If mitigation is an obligation, as the OHCHR and others now argue, logical questions include who specifies levels and adjudicates compliance, and who has recourse to remedy, where, and from whom, when states fail to mitigate adequately? Certainly, mitigation commitments entered under the new UNFCCC agreement are inadequate to prevent widespread harm, and the structure of that agreement does little to institutionalize the OHCHR's understanding of international responsibility (see Chap. 3).

Arguably, an assessment of climate-related harm as obscuring the identification of responsible actors was overly formalistic at the time of the Maldives' and ICC's claims. In light of the unmet need for protection and the increasing legibility of anthropogenic, climate-related harm, the following section revisits the attribution of responsibility amidst the causal complexities enumerated in the HRC report and IACHR hearing.

The Increasing Legibility of Climatic Connections

While indeed "complex," the mechanisms of and contributions to climate-related harm are increasingly well characterized by biophysical and social research ranging from system modeling and emissions accounting, to ecological monitoring and land change mapping, to case studies in affected communities. Such diverse methods and associated findings underpin increasingly confident analyses of attribution common in scientific, policy, journalistic, and political discourse.

In the disclaimer from the OHCHR Report quoted above, three dimensions of socio-ecological complexity are described as legally problematic. These can be classed as aggregative, probabilistic, and prospective aspects of climatic harm.[68] The aggregative aspect refers to the fact that the actions of any one emitter are com-

---

[68] Concerns related to the themes I refer to as aggregation, probabilism, and prospection appear widely in discussions of legal responsibility for climate-related harms, including those considering legal frameworks other than human rights. For further discussion of related challenges in human rights-based approaches in particular, see, for instance, Bodansky (2010a) and Humphreys (2010). Looking beyond human rights-based mobilizations, the Ninth US District Court's dismissal of *Kivalina* vs. *ExxonMobile*, while nominally based in Political Question Doctrine, includes discussion of what I term aggregation and probabilism, as well as issues related to international responsibility discussed here.

bined in their effects on the climate system (and ultimately on places and people) with those of every other emitter; thus the Report's concern about "linking historical greenhouse gas emissions of a particular country" with impacts from climate change. The probabilistic dimension of climatic harm is the subject of the OHCHR's second concern, that "global warming is often one of several factors to climate change-related effects, such as hurricanes, environmental degradation, and water stress." That is, the association of an individual instance of one of these event classes with climatic change (and with anthropogenic climate change more specifically) is probabilistic rather than deterministic. Finally, the prospective aspect of climate harm applies to the OHCHR's concern that "adverse effects of global warming are often projections about future impacts," distinguishing them from the typical judgment about human rights violation, which relies on retrospective analysis.[69]

The first "difficulty" in the OHCHR's list combines aggregative and probabilistic aspects of anthropogenic climate change-related effects, and their rights implications, in a broadly constructed claim against the traceability of harms. Since the rights implications are described elsewhere in the same Report and expanded in subsequent OHCHR statements, I focus here on the aggregative and probabilistic complexities involved in the claim. Rather than "virtually impossible," when taken independently these two aspects of complexity can be productively analyzed. From scientific and economic perspectives then current within policy discussions, atmospherically aggregated contributions to climate harm were already considered readily quantifiable and assignable. In those same discussions, attribution in light of the probabilistic character of climate harm was less well-settled than it is now, through analyses such as those I summarize further down.

By the time of the OHCHR Report methods existed and enjoyed wide currency in climate policy circles for characterizing the "historical greenhouse gas emissions of a particular country" in order to allocate nation-states' responsibilities for anthropogenic climate change. Assuming national emissions are known, in fact, the aggregation of multiple contributions to climate change arguably renders the allocation of responsibility easier. As Knox explained, contrary to the OHCHR's assumption[70]:

> It is not necessary to link the emissions of a particular state to a particular harm in order to assign responsibility for the harm; since all greenhouse gases contribute to climate change, wherever they are released, responsibility could be allocated according to states' shares of global emissions of greenhouse gases...

---

[69] OHCHR 2009, 20.
[70] Knox 2009, 489.

On this basis, it would be possible, at least in principle, to conclude that even if all states contribute to climate change and are therefore joint violators of the human rights affected by it, some states are far more culpable than others, and to allocate responsibility accordingly.

Such allocation methods have long been central to analysts' understanding of the CBDR principle, mentioned by the Maldives and the ICC's legal counsel in the IACHR hearing and now increasingly mobilized in the UNFCCC in the longer analytical formulation of Common but Differentiated Responsibilities and Respective Capabilities in the light of different National Circumstances (CBDR-RC or CBDR-RC-NC). Moreover, precise allocation may be irrelevant from a human rights perspective, as ICC Counsel agued, since any participant in a violation bears responsibility.

In the same analysis of the OHCHR Report, Knox pointed out that historical responsibility and the vast inequalities in per capita emissions at national scale complicate accounts of differential responsibility for globally aggregated emissions. Yet here too, methodologically defensible calculation was not a real issue. Nationally aggregated historical and per capita emissions are available from the UN and other sources, and have played a prominent part, sometimes in combination, within international negotiations as well as political and scholarly analyses of climate justice. More nuanced and flexible methods of numerical analyses are available as well, using, for instance, the Greenhouse Development Rights (GDRs) Framework. GDRs, which appeared in a first formulation in 2004, provide a means by which to transparently allocate burden-sharing based on national levels of responsibility and capability in reference to a threshold level of human development exceeding that of basic needs.[71] More recently, researchers have demonstrated methods for identifying the proportion of a country's emissions causally tied with consumption and capital accumulation in other countries, supporting a yet more economically rigorous, socially precise allocation of responsibility (see Chap. 1).[72]

---

[71] GDRs (Baer et al. 2009) are considered a "framework" because various components of the calculation of responsibility and capacity, including the development threshold, are subject to independent and explicit specification (although the presenting authors designate desirable ranges based on climate science and social welfare). The meaning of responsibility under the GDRs framework is based in part on an operationalization of the concept of climate debt, at national scale. Climate debt designates compensation owed to the poor for damages associated with climate change not caused by them, as well as compensation for the occupation of excessive "atmospheric space"—the capacity of the atmosphere to store greenhouse gases at safe levels—by the wealthy (see Chap. 4 and Roberts and Parks 2009).

[72] Davis and Caldeira 2010; Bergmann 2013.

The probabilistic complexity of climate-related harm incited more reasonable circumspection in the 2009 OHCHR Report, but it too is now tractable to analyses of responsibility. The probabilistic relation between anthropogenic emissions and climate change-related effects still holds, since it extends from the concepts themselves: by definition, "climate" is a statistical abstraction from weather events observed over time within a region.[73] Because weather varies significantly at shorter time scales that those that define climate regimes, no single weather event can be equated with climatic change. Precisely because of the statistical relationship between them, however, recently developed analytical methods and newly accessible data allow scientists to quantify this relatedness, assigning measures of association between specific weather events or patterns and anthropogenic climatic forcing.

A report of the American Meteorological Society, for example, presented such attribution studies of twelve extreme weather events that occurred in 2012.[74] While "not every extreme event can be measurably attributed to climate change," as Bradbury and Tompkins summarized, the majority could[75]:

> For the north-central and northeastern region of the country, the heat wave that resulted in July 2012 being the hottest month on record for the contiguous United States was found to be four times more likely to occur today—as a result of human-induced climate change—than in pre-industrial times. Additionally, due to sea-level rise, extreme flooding along the mid-Atlantic coast on the scale of Hurricane Sandy's impact is more than 30 percent more likely to occur today than it would have been roughly half a century ago. Furthermore, if sea levels at Sandy Hook, NJ were to rise by another 1.2 meters (a scenario projected by the 2013 National Climate Assessment), the flooding level caused by Hurricane Sandy could be expected roughly once every 20 years by the end of the century.

Similar studies were less well-developed when the OHCHR Report was released in 2009. Still, the attribution of impacts to human life from climate change-related events and processes is nothing new. The World Health Organization (WHO), for instance, contributed a study of the health effects of climate change to the first

---

[73] The IPCC (2007b) Glossary states: "Climate in a narrow sense is usually defined as the 'average weather,' or more rigorously, as the statistical description in terms of the mean and variability of relevant quantities over a period ranging from months to thousands or millions of years. The classical period is 30 years, as defined by the World Meteorological Organization (WMO). These quantities are most often surface variables such as temperature, precipitation, and wind. Climate in a wider sense is the state, including a statistical description, of the climate system."

[74] Peterson et al. 2013.

[75] Bradbury and Tompkins 2013.

IPCC report, published in 1990.[76] By 2005, the organization had estimated yearly deaths due to climate change at 150,000.[77] Such tallies depend in part on careful assumptions, of course, and the progression of climate change impact has continued. Accordingly, other more recent studies' methodologies have resulted in far higher estimates of mortality and morbidity.[78]

Many media and civil society organizations readily tie the social costs of disasters to global warming, reflecting the increased likelihood and growing frequency of those events. Journalists and NGOs have drawn attention to climate-related disasters during UN climate negotiations, for example, linking climate change with a deadly landslide in a poor community outside Durban which coincided with the opening of the 2011 meetings in that city, and with Typhoon Haiyan, which struck the Philippines three days prior to the opening of 2013 meetings in Warsaw. Analyses of severe weather events and other disasters, in the public sphere, that link these phenomena with climate change predictions may indicate (and foster) increasing perception of their association.[79]

These quantitative and inferential approaches to probabilistic attribution are possible because climate change is predictable. Using documented climatic conditions from the past, and the record of historical emissions, models of the climate system can predict climate variables up to the present in ways that match remarkably well with observational data. Those successes inspire strong confidence about the same models' abilities to predict the future.[80]

That predictability also underpins the prospective aspect of many claims about climate-related harm, which piqued doubts for some HRC members that were memorialized in the 2009 OHCHR Report. Those doubts, once again, attached to the

---

[76] WHO Task Group 1990.

[77] Patz et al. 2005; WHO 2004, 2005.

[78] For example, the Global Humanitarian Forum's 2009 analysis puts the figure at 300,000 annually, noting more generally that 325 million lives are "seriously affected" by climate change. The Development Assistant Research Associates 2012 report sums deaths due to climate change at 400,000 yearly, and the wider impact of "Our present carbon-intensive energy system and related activities" at 4.5 million lives. See Global Humanitarian Forum 2009; DARA 2012. More recently, the WHO predicted 250,000 deaths annually between 2030 and 2050 (WHO 2014). Haines and Ebi described that estimate as conservative, noting that projected food shortages would likely add 529,000 adult mortalities by 2050 and that 100,000 people could be forced into extreme poverty, with the risk to life that brings, by 2030. Haines and Ebi 2019.

[79] Layzer 2015.

[80] See, for example, Henson 2014, 297–318.

contrast between projections of future climate impact and typical analyses of rights violations, which focus on past actions and impacts. That contrast exemplifies the epistemological contradictions, discussed above, so common between legal reason and scientific understanding of "nature."[81] The apparent obscurity resulting from the two perspectives is a consequence, that is, of the very predictability of climate and environmental systems as opposed, at least in liberal humanist philosophies, to human ones.

Fortunately, at least two approaches exist to facilitate legal protection from prospective harm, with relevance in the climate change context. First, the precautionary principle—which implies the need to protect against possible, but uncertain, dangerous, irreversible, or catastrophic future events—is prominent in both national and international environmental law, and an explicitly recognized principle of the UNFCCC.[82]

Second, in 2012 the Convention took initial steps toward a legal mechanism through which to address "loss and damage" from slow-onset as well as extreme climate change-related events, in effect recognizing the certainty of deleterious human impact as part of the predictability of climate change.[83] Loss and damage has long been seen by some advocates as promising a crucial "third leg" for the Convention, joining mitigation and adaptation to account for the inadequacy of the former and practical limitations on the latter. The potential for robust realization of redress for climate harm through a UNFCCC loss and damage mechanism remains unclear, as contentious debate continues over its terms.[84] Its establishment and subsequent inclusion in the Paris Agreement, albeit in importantly circumscribed form, do however signal a level of acknowledgment in international law of prospective as well as probabilistic forms of attribution: anticipating the social costs of inevitably more extreme weather and increasing ecological instability tied to global warming. The OHCHR also took up the concept in its analyses of climate change and advice to the UNFCCC, arguing that states should "draw upon their human rights obligations" in addressing loss and damage.[85]

Within human rights law, the operational distinction for evaluating prospective dangers is the point at which those dangers can be judged "imminent." In 2015 a Dutch court issued such a judgment for the first time, upholding an NGO's claim

---

[81] Delaney 2003; Herbert et al. 2013.

[82] Martuzzi et al. 2004; UNFCCC 2006.

[83] UNFCCC 2013.

[84] See Mace and Verheyen 2016.

[85] OHCHR 2016.

demanding the national government set stricter emissions reduction targets.[86] Short of this bar, the confidence with which prospective climate harm is now viewed, and the environmental legal approaches established to address it, suggest that at a minimum un- and under-regulated emissions be understood as implying knowing culpability for likely future harm.

With respect to each of these three aspects of complexity that characterize the processes of climate change-related environmental harm, then, the OHCHR's 2009 analysis hewed decisively to a formalist perspective that diverged notably from the assessments then current in influential scientific and policy circles. In subsequent years, the role of human action in climate-related harm has become more legible, and the assignment of responsibility to identifiable actors, including states and corporations, more common, in non-legal analyses as well as some legal ones. It is also worth noting that the accumulating knowledge summarized above does not include more recent contributions from Indigenous communities and traditional knowledges, which document disproportionate cultural and ecological impacts more widespread and further advanced than the ICC could in 2005.[87] Even the OHCHR, however, while strengthening significantly its position on climate change, has not altered its assessment of the claims brought by the Maldives or affirmed those of the ICC, charging major emitters with rights violations from the direct and indirect environmental impacts of anthropogenic climate change. Such lingering lacuna in analyses of global responsibility for climate-related harms bear moral as well as practical implications.

Conclusion

For geographer Jeff Popke, "analytical stances are themselves performative, helping to gather up and constitute the social as a potential site of ethical responsibility and political efficacy."[88] This is surely true for analyses of climate-related harm. Whether activist, Indigenous, scientific, economic, or legal in their character, such analyses impose categorical frames on socio-ecological complexity, defining as they do so the members, relations, and exceptions of a social world.[89] When legal analysis defines relations of causality linking action and harm in terms of unmediated, individual human agency, and interprets obligation to rights bearers as territorially bounded, it

---

[86] *Urgenda Foundation v. The State of the Netherlands.*
[87] See, for example, Wildcat 2013.
[88] Popke 2009, 85.
[89] See also Agamben 1998; Jones 2009; cf. Delaney 2010.

negates responsibility for the most profound, widespread, and disproportionate impacts of anthropogenic climate change, exerting interpretive power in
materially consequential ways across vast swaths of space and time. It is
worth noting, of course, that the authors of country submissions to the HRC
process, compilers of the subsequent OHCHR Report, and IACHR commissioners each assessed the relevance of climate change for human rights under
differing conditions of familiarity with climate science and in fulfillment of
professional roles shaped by distinct mandates. Rather than personal convic-
tions, that is, their interpretative acts reflect institutional positionalities.

In their institutional capacities, the ICC and the Maldives sought to re-
constitute the social with socio-ecologically relational, spatially extensive
analyses of responsibility for climate-related harms. Culturally and geographically specific, their analyses were also simply the first framed in the
formal language of rights to emerge from a growing number of similarly affected and vulnerable communities, who find themselves and their grievances marginalized in climate governance. The fate of such groups seems to
call for legal remedy all the more because it is not, largely, of their own
making.

Yet if the Maldives' or ICC's analysis of anthropogenic climate change as
violating rights internationally were authoritatively affirmed, what then? The
negotiation of divergent interpretations in the 2009 OHCHR Report underscores that law is itself a social construction; its institutional force is backed
by states, who host the adjudication of claims in venues they provide. Where,
then, might claimants seek redress, beyond their own nations, for a violation
of their rights through climate change, and whom might they hold to account? To its credit, the OHCHR has drawn attention to this problem of
venue and the importance of the right to remedy in its recent statements on
climate change.[90] More pithily, legal scholar Stephen Humphreys has suggested that, in the context of climate change, "the law needs a push."[91] To
meet criteria of ethical responsibility or political efficacy, such a push may
need to come from outside.

Legal analyses of climate change and human rights have been elegant,
and influential within limits. Whatever the language linking the two terms, a
strong relation is indeed now "axiomatic" across several discursive commu-

---

[90] For example, OHCHR 2015.
[91] Humphreys 2012.

nities. But activists and scholars of law and social movements have long known that the recognition of rights is not in itself sufficient for their realization. Rather, legal action typically plays a supporting role in real social change, facilitating, if not simply legitimating, other essential conditions. McCann's description of this process includes a prominent role for "rights consciousness" as the outcome of legal action and co-enforcing of collective action.[92] In the case of climate justice, either type of action seems also to require the deepening and broadening of global, socio-ecological consciousness, through non-legal knowledge if not lived experience. A range of mechanisms could be deployed, for instance, by which greater and more even familiarity with emerging scientific understanding of climate change and climate-related harm could be brought to bear on adjudication, with attendant challenges as well as possibilities of translation.[93] Past claimants have often led in incorporating scientific methods of attribution in legal argument, and those indeed represent a leading edge of relevant, research-based knowledge worthy of consideration in realist approaches to causation and responsibility.

Again, however, as McCann's adaptation of "political process" models for legal mobilization studies suggests, changes in more widespread forms of consciousness are crucially related to meaningful legal action.[94] Evolutions in legal reasoning might help foster but would not take the place of a wider social reckoning with the character and magnitude of climate-related harm, and recognition of corresponding human responsibility, particularly within more well-insulated high-emitting polities, like the US, where those uncomfortable realities still sit largely to one side of dominant conceptions of national interest and virtue. Despite official efforts to obscure the long-term impacts of rising emissions,[95] the climate-related shocks of recent years suggest that reckoning is coming. Recognition of responsibility may be further off, and although relevant social movements and leadership are emerging, affirmation of their impulses in legal language could offer additional impetus for collective action.[96]

---

[92] McCann 1994.

[93] See Mertz 2011; Klug and Merry 2016.

[94] McCann's 1994; see McAdam 1982.

[95] Davenport and Landler 2019.

[96] See, for example, Stuart 2019; Graeber 2019.

That is, to borrow Popke's terms once more, legal affirmation of climate rights claims like the ICC's and the Maldives' would furnish at most "a *potential* site of ethical responsibility and political efficacy."[97] Given a confluence of other necessary elements, though, revisiting the legal question of international rights violations from anthropogenic climate change could matter materially. The increasing legibility of the mechanisms that produce climate-related harm, and the increasing scope of those harms, arguably make doing so an ethical responsibility. To analyze global issues of human rights and climate change in purely legal terms, however, would be to perform a kind of mystification: fetishizing the legal in the face of climatic disruptions that challenge some of its foundational categories. In the context of mounting climate-related harm, allowing those categories to limit the purview of legal justice threatens its relevance, along with the lives of innocents.

The Maldives, the OHCHR, and the IACHR each directed continuing effort for the development of rights-based responses to climate change at the United Nations Framework Convention, where leaders of the ICC Petition have also been active. In theory, UN negotiations promise possibilities unavailable in either the IACHR or the HRC. Whereas those institutions looked to established human rights laws and norms to characterize the impacts and production of climate change in legal terms, the UN convention on climate offers a venue for the construction of a purpose-built international regime based in rules and principles and imbued with legal force. As the following chapter explores, however, legality in that regime emerged and evolved within the institutional framework of Westphalian socio-spatial separation as well, and amidst the conflicting efforts of unevenly pitted participants.

Bibliography

Abate, R. S. (2010). Public Nuisance Suits for the Climate Justice Movement: The Right Thing and the Right Time. *Washington Law Review, 85*, 197–252.

ActionAid, & 16 others. (2010a, November). *Climate Justice Briefs #1: Climate Debt*. Retrieved from http://www.ips-dc.org/wp-content/uploads/2010/12/1-Climate-debt.pdf

ActionAid, & 16 others. (2010b, November). *Climate Justice Briefs #12: Human Rights and Climate Justice*. Retrieved from http://www.whatnext.org/resources/Publications/Climate-justice-briefs_full-setA4.pdf

---

[97] Popke 2009, 85, emphasis added.

Agamben, G. (1998). *Homo Sacer*. Redwood City: Stanford University Press.

Atapattu, S. (2015). *Human Rights Approaches to Climate Change: Challenges and Opportunities*. London: Routledge.

Averill, M. (2010). Getting into Court: Standing, Political Questions, and Climate Tort Claims. *Review of European Community & International Environmental Law, 19*(1), 122–126.

Baer, P., Athanasiou, T., Kartha, S., & Kemp-Benedict, E. (2009). Greenhouse Development Rights: A Proposal for a Fair Global Climate Treaty. *Ethics, Place & Environment, 12*(3), 267–281.

Bell, D. A., Jr. (1980). Brown v. Board of Education and the Interest-Convergence Dilemma. *Harvard Law Review, 93*(3), 518–533.

Bergmann, L. (2013). Bound by Chains of Carbon: Ecological–Economic Geographies of Globalization. *Annals of the Association of American Geographers, 103*(6), 1348–1370.

Bix, B. H. (2005). Law as an Autonomous Discipline. In P. Cane & M. Tushnet (Eds.), *The Oxford Handbook of Legal Studies* (pp. 975–987). Oxford/New York: Oxford University Press.

Blomley, N. (1994). *Law, Space, and the Geographies of Power*. New York: Guilford.

Blomley, N. (2008). Simplification Is Complicated: Property, Nature, and the Rivers of Law. *Environment and Planning A, 40*(8), 1825–1842.

Blomley, N., Delaney, D., & Ford, R. (Eds.). (2001). *The Legal Geographies Reader: Law, Power and Space*. Hoboken: Wiley.

Bodansky, D. (2010a). Climate Change and Human Rights: Unpacking the Issues. *Georgia Journal of International and Comparative Law, 38*, 511–524.

Bond, P. (2011). *The Politics of Climate Justice*. London: Verso and Pietermaritzburg: University of KwaZulu-Natal Press.

Bradbury, J., & Tompkins, C. F. (2013). *New Report Connects 2012 Extreme Weather Events to Human-Caused Climate Change* [online]. World Resources Institute. Available from: http://www.wri.org/blog/2013/09/new-report-connects-2012-extreme-weather-events-human-caused-climate-change

Braverman, I., et al. (Eds.). (2014). *The Expanding Spaces of Law: A Timely Legal Geography*. Redwood City: Stanford University Press.

Burger, M., & Wentz, J. A. (2015). *Climate Change and Human Rights* [online]. Columbia University Academic Commons. Available from: https://doi.org/10.7916/D8PG1RRD.

Castree, N. (2005). *Nature*. Abingdon/New York: Routledge.

Chapman, M. (2010). Climate Change and the Regional Human Rights Systems. *Sustainable Development Law & Policy, 10*(2), 37–38.

Cover, R. (1983). Nomos and Narrative. *Harvard Law Review, 97*(4), 4–68.

DARA (Development Assistant Research Associates). (2012). *Climate Vulnerability Monitor, 2nd edition: A Guide to the Cold Calculus of a Hot Planet*. Retrieved from http://daraint.org/climate-vulnerability-monitor/climate-vulnerability-monitor-2012/

Davenport, C., & Landler, M. (2019). Trump Administration Hardens Its Attack on Climate Science. *New York Times* [online], 27 May. Available from: https://www.nytimes.com/2019/05/27/us/politics/trump-climate-science.html

Davis, S. J., & Caldeira, K. (2010). Consumption-Based Accounting of CO_2 Emissions. *Proceedings of the National Academy of Sciences, 107*(12), 5687–5692.

Delaney, D. (2003). *Law and Nature*. Cambridge: Cambridge University Press.

Delaney, D. (2010). *The Spatial, the Legal and the Pragmatics of World-Making: Nomospheric Investigations*. Abingdon: Routledge.

Derman, B. B. (2013). Contesting Climate Injustice During COP17. *South African Journal on Human Rights: Climate Change Justice: Narratives, Rights and the Poor, 29*(1), 170–179.

Derman, B. B. (2014). Climate Governance, Justice, and Transnational Civil Society. *Climate Policy, 14*(1), 23–41.

Dulitzky, A. (2006, November 16). Letter to Paul Crowley. *New York Times.* Retrieved from http://graphics8.nytimes.com/packages/pdf/science/16commissionletter.pdf

Galanter, M. (1974). Why the "Haves" Come Out Ahead: Speculations on the Limits of Legal Change. *Law & Society Review, 9*(1), 95–160.

Global Humanitarian Forum. (2009). *The Anatomy of a Silent Crisis* (Human Impact Report—Climate Change). Retrieved from http://www.ghf-ge.org/human-impact-report.php

Gloppen, S., & St. Clair, A. (2012). Climate Change Lawfare. *Social Research: An International Quarterly, 79*(4), 899–930.

Government of the Maldives. (2008, September 25). *Submission of the Maldives to the OHCHR Under Human Rights Council Res. 7/23.* Retrieved from http://www.ohchr.org/english/issues/climatechange/docs/submissions/Maldives_Submission.pdf

Government of the United States. (2008, September 25). *Submission of the United States to the OHCHR Under Human Rights Council Res. 7/23.* Retrieved from http://www2.ohchr.org/english/issues/climatechange/docs/submissions/USA.pdf

Graeber, D. (2019). If Politicians Can't Face Climate Change, Extinction Rebellion Will. *New York Times*, [online], 1 May. Available from: https://www.nytimes.com/2019/05/01/opinion/extinction-rebellion-climate-change.html. Accessed 30 May 2019.

Haines, A., & Ebi, K. (2019). The Imperative for Climate Action to Protect Health. *New England Journal of Medicine, 380*(3), 263–273. https://doi.org/10.1056/NEJMra1807873.

Henson, R. (2014). *The Thinking Person's Guide to Climate Change.* Boston: American Meteorological Society.

Herbert, S. K. (1997). *Policing Space: Territoriality and the Los Angeles Police Department.* Minneapolis: University of Minnesota Press.

Herbert, S., Derman, B., & Grobelski, T. (2013). The Regulation of Environmental Space. *Annual Review of Law and Social Science* [online], *9*(1), 227–247. Available from: https://doi.org/10.1146/annurev-lawsocsci-102612-134034. Accessed 30 May 2019.

HRC (U.N. Human Rights Council). (2009, March 31). *Human Rights Council Resolution 10/4 Human Rights and Climate Change.* U.N. Symbol A/HRC/RES/10/4. Geneva. Retrieved from http://ap.ohchr.org/documents/E/HRC/resolutions/A_HRC_RES_10_4.pdf

Humphreys, S. (2010). *Human Rights and Climate Change.* Cambridge: Cambridge University Press.

Humphreys, S. (2012). Climate Change and Human Rights: Where Is the Law? In *Address Delivered at the 2012 Rafto Prize Ceremony.* Bergen: Copy on file with the author.

ICC (Inuit Circumpolar Council). (2005). *Petition to the Inter American Commission on Human Rights Seeking Relief from Violations Resulting from Global Warming Caused by Acts and Omissions of the United States.* ICC Canada. Available from: http://www.inuitcircumpolar.com/inuit-petition-inter-american-commission-on-human-rights-to-oppose-climate-change-caused-by-the-united-states-of-america.html

IPCC (Intergovernmental Panel on Climate Change). (2007b). *AR4 SYR Synthesis Report Annexes—Glossary A–D.* Retrieved February 20, 2015, from http://www.ipcc.ch/publications_and_data/ar4/syr/en/annexessglossary-a-d.html

Jones, R. (2009). Categories, Borders and Boundaries. *Progress in Human Geography, 33*(2), 174–189.

Klug, H., & Merry, S. E. (Eds.). (2016). *The New Legal Realism. Vol. 2: Studying Law Globally*. Cambridge: Cambridge University Press.

Knox, J. H. (2009). Linking Human Rights and Climate Change at the United Nations. *Harvard Environmental Law Review, 33*, 477–498.

Layzer, J. A. (2015). *The Environmental Case: Translating Values into Policy*. Thousand Oaks: Sage.

Lesnikowski, A., et al. (2016). What Does the Paris Agreement Mean for Adaptation? *Climate Policy* [online], *17*(7), 825–831. Available from https://doi.org/10.1080/14693062. 2016.1248889. Accessed 30 May 2019.

Limon, M. (2009). Human Rights and Climate Change: Constructing a Case for Political Action. *Harvard Environmental Law Review, 33*, 439–476.

Mace, M. J., & Verheyen, R. (2016). Loss, Damage and Responsibility After COP21: All Options Open for the Paris Agreement. *Review of European, Comparative & International Environmental Law* [online], *25*(5), 197–214. Available from: https://doi.org/10.1111/reel.12172. Accessed 30 May 2019.

Martuzzi, M., et al. (2004). *The Precautionary Principle: Protecting Public Health, the Environment and the Future of Our Children*. WHO Regional Office for Europe Copenhagen. Available from: http://www.asser.nl/media/2227/cms_eel_96_1_book-precautionary-principle-protecting-public-health-the-environment.pdf

Mary Robinson Foundation—Climate Justice. (2017). *Principles of Climate Justice*. Available from: https://www.mrfcj.org/principles-of-climate-justice/

Mayer, B. (2016). Human Rights in the Paris Agreement. *Climate Law, 6*(1–2), 109–117.

McAdam, D. (1982). *Political Process and Black Insurgency, 1930–1970*. Chicago: University of Chicago Press.

McCann, M. W. (1994). *Rights at Work: Pay Equity Reform and the Politics of Legal Mobilization*. Chicago: University of Chicago Press.

McInerney-Lankford, S., Darrow, M., & Rajamani, L. (2011). *Human Rights and Climate Change: A Review of the International Legal Dimensions*. Washington, DC: The World Bank.

Merry, S. E. (2006). *Human Rights and Gender Violence: Translating International Law into Local Justice*. Chicago: University of Chicago Press.

Mertz, E. (2011). Undervaluing Indeterminacy: Translating Social Science into Law. *DePaul Law Review* [online], *60*, 397–412. Available from: http://via.library.depaul.edu/law-review/vol60/iss2/7. Accessed 30 May 2019.

Mertz, E., et al. (Eds.). (2016). *The New Legal Realism: Volume 1: Translating Law-and-Society for Today's Legal Practice*. New York: Cambridge University Press.

Ninth International Conference of American States. (1948). *American Declaration on the Rights and Duties of Man*. Bogota, Colombia (March 30–May 2). Available from: https://www.cidh.oas.org/Basicos/English/Basic2.American%20Declaration.htm

OAS (Organization of American States). (2007). *Human Rights and Global Warming*. IACHR Hearings and Other Public Events. Available from: http://www.oas.org/es/cidh/audiencias/Hearings.aspx?Lang=en&Session=14

OAS (Organization of American States). (2008). *Human Rights and Climate Change in the Americas*. Available from: https://www.oas.org/dil/AGRES_2429.doc

OAS (Organization of American States). (2011). *Annual Report of the Inter-American Commission on Human Rights*. Available from: http://www.oas.org/en/iachr/docs/annual/2011/toc.asp

OAS (Organization of American States). (2015). *IACHR Expresses Concern Regarding Effects of Climate Change on Human Rights*. Available from: http://www.oas.org/en/iachr/media_center/preleases/2015/140.asp

OHCHR (Office of the United Nations High Commissioner for Human Rights). (2009). *Report of the Office of the United Nations High Commissioner for Human Rights on the Relationship Between Climate Change and Human Rights*. U.N. Symbol A/HRC/10/61.

OHCHR (Office of the United Nations High Commissioner for Human Rights). (2015). *Understanding Human Rights and Climate Change (Submission to the 21st Conference of Parties to the UNFCCC)*. Geneva, November 27. Available from: http://www.ohchr.org/Documents/Issues/ClimateChange/COP21.pdf

Osofsky, H. M. (2006). The Inuit Petition as a Bridge? Beyond Dialectics of Climate Change and Indigenous Peoples' Rights. *American Indian Law Review, 31*(2), 675–697.

Patz, J. A., Campbell-Lendrum, D., Holloway, T., & Foley, J. A. (2005). Impact of Regional Climate Change on Human Health. *Nature, 438*(7066), 310–317.

Peterson, T. C., Alexander, L. V., Allen, M. R., Anel, J. A., Barriopedro, D., Black, M. T., … others. (2013). Explaining Extreme Events of 2012 from a Climate Perspective. *Bulletin of the American Meteorological Society, 94*(9), S1–S74.

Popke, J. (2009). Geography and Ethics: Non-Representational Encounters, Collective Responsibility and Economic Difference. *Progress in Human Geography* [online], *33*(1), 81–90. Available from: https://doi.org/10.1177%2F0309132508090441. Accessed 30 May 2019.

Rajamani, L. (2010). The Increasing Currency and Relevance of Rights-Based Perspectives in the International Negotiations on Climate Change. *Journal of Environmental Law, 22*(3), 391–429. https://doi.org/10.1093/jel/eqq020.

Report of the Office of the United Nations High Commissioner for Human Rights on the Relationship Between Climate Change and Human Rights. (2009). (A/HRC/10/61) [online]. UN Human Rights Council, 15 January. Available from: https://documents-ddsny.un.org/doc/UNDOC/GEN/G09/103/44/PDF/G0910344.pdf?OpenElement. Accessed 30 May 2019.

Robbins, P. (2007). *Lawn People: How Grasses, Weeds, and Chemicals Make Us Who We Are*. Philadelphia: Temple University Press.

Robbins, P. (2012). *Political Ecology: A Critical Introduction* (Vol. 20). Chichester: John Wiley & Sons.

Roberts, J. T., & Parks, B. C. (2009). Ecologically Unequal Exchange, Ecological Debt, and Climate Justice: The History and Implications of Three Related Ideas for a New Social Movement. *International Journal of Comparative Sociology, 50*(3–4), 385–409.

Rogelj, J., et al. (2016). Paris Agreement Climate Proposals Need a Boost to Keep Warming Well Below 2 °C. *Nature, 534*(7609), 631–639.

Sachs, W. (2008). Climate Change and Human Rights. *Development, 51*(3), 332–337.

Scheingold, S. A. (1974). *The Politics of Rights*. New Haven: Yale University Press.

Shearer, C. (2011). *Kivalina: A Climate Change Story*. Chicago: Haymarket Books.

Stuart, T. (2019). Sunrise Movement, the Force Behind the Green New Deal, Ramps Up Plans for 2020. *Rolling Stone* [online], 1 May. Available from: https://www.rollingstone.com/politics/politics-features/sunrise-movement-green-new-deal-2020-828766/. Accessed 30 May 2019.

Submission of the Maldives to the Office of the UN High Commissioner for Human Rights. (2008). *Human Rights Council Resolution 7/23. "Human Rights and Climate Change"* [online]. 25 September 2008. Available from: https://www.ohchr.org/Documents/Issues/ ClimateChange/Submissions/Maldives_Submission.pdf. Accessed 10 Nov 2017.

Swyngedouw, E. (1999). Modernity and Hybridity: Nature, Regeneracionismo, and the Production of the Spanish Waterscape, 1890–1930. *Annals of the Association of American Geographers, 89*(3), 443–465. https://doi.org/10.1111/0004-5608.00157.

Turk, A. T. (1976). Law as a Weapon in Social Conflict. *Social Problems, 23*(3), 276–291.

UNFCCC (United Nations Framework Convention on Climate Change). (2006). *United Nations Framework Convention on Climate Change Handbook.* Bonn: Intergovernmental and Legal Affairs, Climate Change Secretariat. Available from: https://unfccc.int/re-source/docs/publications/handbook.pdf

UNFCCC (United Nations Framework Convention on Climate Change). (2009). *Copenhagen Accord.* U.N. Symbol FCCC/CP/2009/L.7.

UNFCCC (United Nations Framework Convention on Climate Change). (2013). Report of the Conference of the Parties on Its Nineteenth Session, Held in Warsaw from 11 to 23 November 2013. Addendum. Part Two: Action Taken by the Conference of the Parties at Its Nineteenth Session. U.N. Symbol FCCC/CP/2013/10/Add.1.

United Nations. (1992). *United Nations Framework Convention on Climate Change.* U.N. Symbol A/AC.237/18.

Werksman, J. (2011). *The Challenge of Legal Form at the Durban Climate Talks.* World Resources Institute. Available from: http://www.wri.org/blog/2011/11/challenge-legal-form-durban-climate-talks

WHO (World Health Organization). (2004). *Comparative Quantification of Health Risks: Global and Regional Burden of Disease Attributable to Selected Major Risk Factors.* Geneva: World Health Organization. Retrieved from http://www.who.int/healthinfo/ global_burden_disease/cra/en/

WHO (World Health Organization). (2005). *Deaths from Climate Change* (map). Available at: http://www.who.int/heli/risks/climate/climatechange/en/

WHO (World Health Organization). (2014). *Quantitative Risk Assessment of the Effects of Climate Change on Selected Causes of Death, 2030s and 2050s.* Geneva: World Health Organization.

WHO Task Group. (1990). *Potential Health Effects of Climatic Change.* WHO Symbol WHO/PEP/90/10. Geneva: World Health Organization.

Wildcat, D. R. (2013). Introduction: Climate Change and Indigenous Peoples of the USA. *Climatic Change, 120*(3), 509–515. https://doi.org/10.1007/s10584-013-0849-6.

Zemans, F. K. (1983). Legal Mobilization: The Neglected Role of the Law in the Political System. *The American Political Science Review, 77*(3), 690–703.

Legal Cases

Petition Requesting for Investigation of the Responsibility of the Carbon Majors for Human Rights Violations or Threats of Violations Resulting from the Impacts of Climate Change. CHR-NI-2016-0001 (Republic of the Philippines Commission on Human Rights, Quezon City. 2016).

Urgenda Foundation v. The State of the Netherlands (Ministry of Infrastructure and the Environment), C/09/456689 / HA ZA 13–1396 (Dist. Ct., The Hague. 2015).

Law, Power, and the COPs

3

The United Nations Framework Convention on Climate Change (UNFCCC) is a purpose-built, rules-based international treaty designed to address the risks of anthropogenic climate change. The foundational rules, structures, and principles of the UNFCCC partly reflect the known, correlated differences between countries in their economic development, responsibility for, and vulnerability to climate change. Such principles have therefore provided the grounds for much of climate justice advocacy in the international context. Indeed, for many, the governmental and socio-spatial separations of the international system, which structure forums like the UNFCCC, imply the necessity of international legal rules and principles with which to mediate global challenges like the shared burdens of climate change impact and regulation.

However, because national sovereignty is enshrined within the international system, because deep inequalities persist between countries, and because they are produced by negotiations, those rules have always also been constructed through the give and take of power between countries. That is, at the climate convention, legality exists simultaneously, in juxtaposition, and co-constitutively with geo-economic power politics. One example from the earlier history of the convention is the influence the US exerted over the construction of specific aspects of the Kyoto Protocol (KP), which it then did not ratify.[1] More generally, because of its outsized

---

[1] Patterson 2011.

© The Author(s) 2020
B. B. Derman, *Struggles for Climate Justice*,
https://doi.org/10.1007/978-3-030-27965-3_3

influence as a major emitter and economic power, the US has been able to operate from a position on the periphery of the legal framework of the UNFCCC to shape that framework in the service of its own, nationally defined interest, all too often limiting the potential of international cooperation along the way.

The restructuring of the "legal form" of the convention between Copenhagen and Paris illustrates that dynamic, which, as before, came to profoundly shape the next major phase of the international climate governance regime. In this episode, the US held out the promise of a legal agreement as a way of pushing its own vision for the relations between countries that the new framework would encode. It was able to do so because the less powerful countries and their civil society allies had little to rely on but the rules and legal structures of the convention, using them both as a means of forwarding their own interests within the negotiations and as a potent language for the cause of global justice.[2] Participating in the legality of the UNFCCC is optional, instead, for powerful parties like the US, while domestic forces often drive their ability to do so constructively. This chapter shows how legalistic arguments—and the conditional acceptance of the international rule of law itself—offered modalities for geo-economic contest as it played out in the re-design of the international climate regime, including its ambitions and means for quelling the progression and ameliorating the worst outcomes of anthropogenic warming. In tracking these dynamics, I draw on a combination of official documents, position papers, observation at COPs, and interviews with some of the country and civil society delegates intimately involved.

Following the Paris Agreement, US President Donald Trump considered even a legal framework forcefully shaped by the US too great a check on sovereign economic interests, while withdrawing from that agreement offered opportunity for narrow domestic political gain. As the new US administration distanced itself from the arrangements for which the previous one had fought, casting legal multilateralism in climate governance in doubt once again, still more uncertainty opened for the future of the climate and those most affected by its accelerating alteration.

Mobilizing Climate Legality in the UNFCCC

The UNFCCC emerged from the 1992 UN Conference on Sustainable Development, codifying participating countries' collective wish "to achieve ... stabilization of greenhouse gas concentrations in the atmosphere at a level that would prevent

---

[2] Goodale and Merry 2007.

dangerous anthropogenic interference with the climate system."[3] The convention is a treaty which provides a framework of rules to support progress toward that aim through negotiated consensus among delegations representing each signatory nation, in accordance with a set of agreed principles, and with the consultative input of non-state "observers."[4] Over time, subsidiary bodies have emerged devoted to specific tasks under the Convention, which meet throughout the year. The Conference of the Parties (COP) convenes once yearly to approve and consolidate the work of those bodies and set their future direction. Until recently, the UNFCCC's most widely touted achievement has been the Kyoto Protocol (KP), agreed in 1997. The KP encoded the only legally binding national targets for climate change mitigation to enter force thus far and launched enduring transborder financial instruments to facilitate achieving those targets without radically altering developed countries' economies or energy infrastructures.[5] In 2015, parties to the convention negotiated a second landmark deal, the Paris Agreement, which laid the cornerstones of a new regime intended to have similar legal status beginning in 2020. Key transformations occurred, however, in conceptions of responsibility between Kyoto and Paris as legal rules and the felt importance of legality itself were mobilized both by and against those seeking climate justice.

As the original Kyoto commitments' planned sunset date of 2012 approached and the parties struggled to draft a follow-on agreement, negotiations became

---

[3] United Nations 1992. In its entirety, Article 1 states: "The ultimate objective of this Convention and any related legal instruments that the Conference of the Parties may adopt is to achieve, in accordance with the relevant provisions of the Convention, stabilization of greenhouse gas concentrations in the atmosphere at a level that would prevent dangerous anthropogenic interference with the climate system. Such a level should be achieved within a timeframe sufficient to allow ecosystems to adapt naturally to climate change, to ensure that food production is not threatened and to enable economic development to proceed in a sustainable manner."

[4] See Arts 1998; Betsill and Corell 2008 on the latter.

[5] UNFCCC 1997. The KP, agreed in 1997, entered into force in 2005. It binds "developed" countries (excluding the US) to emissions reduction targets. Among the framework instruments it created is the Clean Development Mechanism (CDM), a system of financial transfers from developed countries to fund energy and economic development projects in developing countries, as a means of meeting developed countries' obligations. Though the CDM has been considered by some parties to be an important model for future agreements, its efficacy in achieving mitigation or sustainable development in developing countries is much contested. Chapter 4 touches on critiques of the CDM and related exchange-based mechanisms for mitigation in terms of their justice implications and underlying relations. On KP financial instruments, see also Paterson 2011.

slower and more contentious.[6] The tensions which erupted in 2009 at COP15, described in the Introduction, inaugurated a period of institutional crisis. As it happened, the treaty was quickly rehabilitated in dramatically altered form, which would be enshrined in the Paris Agreements. In the interim, thorny issues of international justice rooted in uneven development and differential responsibility for climate change dogged the COPs, as negotiating parties and other institutional players sought an elusive new deal that could be politically palatable to major emitters, protect the most vulnerable states and communities, and shift the global trajectory of rising greenhouse gas emissions.[7]

COP meetings encompass a variety of formal and informal interactions,[8] but because the Convention is constructed and participation performed through legalistic language and procedures, both formal proceedings and surrounding political debates often involve the invocation of legal norms and obligations. If parties' and observers' interventions frequently take rhetorical shape in a universalizing language of rights, responsibilities, and principles, however, the individual economic interests of states and multi-state negotiating blocs often lie readily apparent beneath them.

As geographer Neil Smith showed for the US influence in the formation of the UN and the League of Nations before it, the "global" visions around which international institutions are constructed can encode the partial perspectives of powerful participants.[9] Global governance, that is, can institutionalize, *in the name of the global*, the enabling means for narrower imperial ambitions. Those means, however, stand in complex relation to the universalist and cosmopolitan discourses that frame them, which can both reinforce and challenge the hegemony of powerful actors.[10] The UNFCCC's founding principles offer such malleable discursive resources: "equity" and Common but Differentiated Responsibilities (CBDR) in particular.[11] Both principles lack definition in their original articulation. Their political salience inheres in their potential to specify a variety of relations among parties to the Convention, just as the indeterminacy of many legal principles facilitates their deployment in political struggle.[12]

---

[6] See, for example, Grubb 2011.

[7] See, for instance, Klinsky and Winkler 2014.

[8] Derman 2014.

[9] Smith 2003.

[10] Minow 1991; James 2001; Mitchell 2011.

[11] These are, politically, the key terms of Article 3, "Principles" in United Nations 1992, discussed below. As discussed in Chap. 2, CBDR is increasingly invoked in expanded forms drawn from the Convention and now the Paris Agreements.

[12] Galanter 1974; Herbert et al. 2013.

Debate over the meanings of equity and CBDR have been fraught at the UNFCCC because its negotiations pit unequally resourced parties against each other in formally *equal* negotiation over regulatory commitments, with potentially dramatic consequences for economic stability and development.[13] For many parties and observers, the inclusion of equity and CBDR in the Convention suggests that those regulatory commitments should respond to inequalities in the human production and impacts of climate change, as well as differing levels of development. Thus, like the debates over the human rights implications of climate change examined in Chap. 2, facially formal wrangling within the UNFCCC over the meanings of CBDR and equity are ultimately contests about the recognition of socio-spatial ties of responsibility for climate change-related harm and vulnerability to it. These factors of concern are themselves co-produced because of the correlation between development and historical emissions, on the one hand, and vulnerability and lack of development on the other (see Chap. 1).

Despite dubious potential for enforcement, then, international law has provided a prominent discursive terrain for geo-economic and climate justice struggles at the UNFCCC. This terrain was actively reconstructed, largely in favor of more economically powerful negotiating parties, as the successor to Kyoto's first legally binding commitment period was hammered out. In the years following COP15, the US and other developed countries combined the conditional promise of a legally binding treaty with the exercise of their geo-economic power to reconfigure the grounds of debate over a future climate treaty and the shape it would ultimately take: first, by moving discussion of mitigation commitments between legal and non-legal registers, and second, by seeking to control the meanings and downplay the importance of equity and CBDR.

Reneging in Copenhagen

Several interviewees and other analysts located a key transitional moment in the evolution of the UNFCCC at the 2009 meetings in Copenhagen (COP15). There, party delegations, many represented by heads of state, faced head-on the challenge of drafting of a sequel to the KP's first commitment period, under the glare of unusually intense media attention. The previous years had been frenetic for climate politics and policy, encompassing increased popular attention,[14] new

---

[13] This is Steve Gardiner's concern over the potential of existing institutions to resolve the climate political conundrum. See Gardiner 2011.

[14] For example, Gore 2006; Kolbert 2006; Lynas 2007.

national, subnational, and municipal efforts,[15] and the rapid development of the apparent bases for a new global agreement. To wit, the parties had agreed at the Bali COP in 2007 to structure future agreements through a "two track system." This system would retain but also supplement the KP's original legal form, under which the principle of CBDR was operationalized in the division of country parties into Annexes: developed countries, enumerated in "Annex I," received legally binding commitments to reduce their emissions, while developing, or "non-Annex I," countries were assigned no commitments but expected to "graduate" to Annex I status upon achieving development goals. The two tracks of Bali meant that the KP arrangement would continue in a second commitment period, while a parallel process would emerge and enroll developing countries as well as the US (which had not joined the KP and had thus taken on no mitigation commitments under the Convention) in new internationally coordinated mitigation efforts.[16]

By Copenhagen, however, developed countries and other wealthy member states faced a deepening economic recession, and climate change legislation had foundered in the US. As the COP opened, developed country delegations encountered those of poor developing countries and an unprecedented number of observers and activists[17] allied, roughly, in global campaigns for climate justice. Their demands included massive financial transfers to the Global South, and ambitious emissions cuts in the North.[18] After two weeks rife with procedural haggling, protest, and accusations of foul play, the summit ended with the Copenhagen Accord: a two-page "politically-binding" re-statement of intention, inviting all countries' voluntary pledges to reduce emissions in lieu of a legally binding agreement for specific emissions targets allocated according to CBDR, in the style of the KP. Just as significant, for many observers, was the fact that the Accord emerged from a drafting process held in parallel with the main negotiations, which involved only a few key parties and excluded the larger majority.[19] Finally, in 2010, WikiLeaks cables exposed developed nation delegates' candid

---

[15] For example, Mayors Climate Protection Center 2005; *Massachusetts v. EPA*; Cole 2007; C40 Cities Climate Leadership Group 2015.

[16] See UNFCCC 2007. The two tracks of the "Bali Road Map" therefore respected the KP's central commitment to CBDR, while seeking to address the major weaknesses in its structure: laying the groundwork for inclusion of the US and China, the world's top emitters, as well as other developing countries in a more encompassing international regime.

[17] Fisher 2010.

[18] Observers' and activists' climate justice demands are examined further in Chap. 4. See also Bond 2011.

[19] For example, Khor 2010a.

skepticism about the potential of established negotiating principles and practices under the UNFCCC, as well as the hardball economic tactics employed by the US to gain recognition of the Accord from developing countries.[20] Reception for the Copenhagen Accord was decidedly mixed. The American administration, its negotiating allies, and a minority of sympathetic observers hailed it as a welcome new beginning.[21] For many other parties and observers, by contrast, the Accord represented a crisis of efficacy and legitimacy for the Convention and global climate governance as a whole. To the disappointment of many in the climate justice community (though unsurprisingly, given its brevity), the Accord made no mention of human rights. One analyst of climate law and politics with experience from state, intergovernmental, and NGO roles described the Copenhagen outcome to me as the end of an era, signaling the demise of a theory of change that had emphasized consensus and coordination:

[T]he last window of opportunity opened around 2006–2007.... You had Gore's movie ["*An Inconvenient Truth*"] ... the 4th assessment report of the IPCC ... the UNDP Human Development report of 2007 ... the Stern review ... you could see many, many different communities pitching in....

That led initially to the [2007] COP in Bali ... that pivotal moment when the US was forced to join in the consensus and agree to the Bali action plans.... You can also see around that time a lot of domestic governments stepping up to take action ... [and widespread] measures at the subnational level ....

And all of that builds towards momentum in Copenhagen. And then everything collapsed. The Copenhagen accord was a face saver more than anything else, but everything clearly, clearly came to a head in Copenhagen and people realized that what they had been pushing for, this sort of "silver bullet" moment wasn't happening.

Severely undercutting hope for ambitious mitigation efforts, the Accord was wildly disappointing to environmentalists around the globe. For Convention insiders, however, it was also remarkable for the withdrawal it seemed to represent from any recognizable legal framework for international climate change response. Moreover, it indexed once again the influence held by the US over the trajectory of the Convention, President Barack Obama reportedly having personally drafted some sections of the Accord. Thus, another interviewee, a long-time UNFCCC observer and theorist of equity in climate governance, quipped, "my view of what happened

---

[20] See Carrington 2010; Lister 2010.

[21] For example, Bodansky 2010.

[in Copenhagen] is that the US decided to renege on the Convention." As it happened, though, reneging in Copenhagen shaded into renegotiating some of the central tenets of the international climate regime, including the role and relevance of legality within that regime, against a stark backdrop of unequal power between countries.

Conventional Commitments and Climatic Futures

Despite the rupture initiated in Copenhagen—indeed *because* of the threat it posed to established norms, ambition, and therefore to the legitimacy and potential of the UNFCCC—the evolution of the convention would quickly return to a more recognizable legal-multilateral register. The influence of the interests that shaped the Accord, however, has proven more lasting.

On the heels of what the secretariat and many parties and observers experienced as failure in Copenhagen,[22] energetic preparations began for the 2010 COP16, to be held in Cancun. Although conflict and confrontation would again permeate activity around the COP—and, to a lesser extent, within it—the treaty body would renew its commitment to legalistic multilateralism there, closing the conference with a raft of newly agreed texts. Along with numerous details related to forestry; the Monitoring, Reporting and Verification of developing countries' domestic mitigation efforts (MRV in UNFCCC discourse); and plans for a new adaptation funding mechanism dubbed the Green Climate Fund (GCF), parties officially recognized HRC Resolution 10/4. The Cancun Agreements port language directly from that document, in the first acknowledgment of the human rights implications of climate change to emerge from the UNFCCC, although they forego specifying what those implications could require of signatories.[23]

The Cancun Agreements can therefore be read as a return to form, of sorts, for the convention. They can also be seen, however, as a transposition of the impulses underpinning the Copenhagen Accord—the shift from legally binding *commitments* to a voluntary or "pledge and review" approach, and less differentiation in the obligations of developing and developed countries—into the traditional legal-multilateral register. The Cancun Agreements also signaled a new definition for key procedural norms that had long helped to constitute that register, as a means of producing texts that could be seen as putting the Convention back on track.

---

[22] Vidal 2010; Schueneman 2010.
[23] CIEL 2011; Johl and Duyck 2012.

Legalizing Copenhagen's Right Turn in Cancun

As the last interviewee quoted observed, "if you look at the Cancun Agreements, they're almost the consensus expression of the Copenhagen Accord." That is, key aspects of the Accord—recognition of countries' voluntary pledges rather than negotiated targets to reduce GHG emissions, and the erosion of the "Annex" system through which the KP had operationalized CBDR—were now included in formally agreed texts under the Convention, rather than merely "taken note of," as they had been by parties acting independently in Copenhagen. The shift these new agreements embodied from the two "tracks" laid out in Bali was dramatic, and demanded explanation. To journalist and director of the intergovernmental organization the South Center, Martin Khor, negotiations in Cancun "seemed ... to revolve around meeting the requirements of the most powerful country, so that very modest progress could be allowed to be made in other areas, and that Cancun would thus be saved from being termed a total failure."[24]

Indeed, the parties' adoption and expansion in Cancun of directions laid out in Copenhagen suggest that what observers had classed as the "hardline," "all-or-nothing" strategy pursued by the US paid off, in apparently more widely accepted agreements under the official seal of the treaty body.[25] Given that these entailed increased obligations for developing countries and decreasing ones for developed ones, it is difficult to explain their adoption without reference to political calculation, by which small gains, fear of blame for the potential collapse of the convention, and the possibility of ancillary benefits outweighed greater ambition and commitment to earlier protections in the face of developed countries' uncompromising stance.[26] In this sense, moving outside of the established legal framework for coordinated climate response in Copenhagen—and the institutional crisis that move instigated—facilitated shifting the grounds of international debates, which then continued within the legal-multilateral register in Cancun and afterward.

At the same time, the multilateralism of climate negotiations itself emerged altered, or perhaps clarified, from the Cancun COP. First, adoption of the Cancun Agreements occurred against the objection of the Bolivian delegation, which

---

[24] Khor 2010a, 11.

[25] Goldenberg 2010b; Walsh 2010.

[26] Khor 2010a, 12. Khor offers detailed analysis of the Cancun Agreements as reflecting the protracted debates and political calculations of a bifurcated negotiating landscape where developed and developing countries' interests are rarely seen to coincide. Paterson 2011 characterizes this dynamic as a determining one in the longer trajectory of the UNFCCC.

remained opposed to the procedural origins as well as key substantive elements within the texts, including the non-binding character and low ambition of developed country pledges.[27] While the UNFCCC had never formally specified the meaning of its "consensus" decision policy, adoption in the face of the formal, sustained objection by a negotiating party was novel.[28] Lack of support or recognition for Bolivia's rejection of the Agreements underscored the broadly felt need to legitimate institutional efforts in Cancun, despite the meager promise of mitigation and the evolving understanding of CBDR to which they gave rise.

Second, while the Cancun Agreements were thereby formally adopted with broad if not truly consensual approval, their genesis lay in more narrowly inclusive forms of multilateralism reminiscent of those out of which the Copenhagen Accord had been born. Critics likened Cancun's small group drafting sessions and one-on-ones between the secretariat and influential parties to practices common in the World Trade Organization (WTO), contrasting them with the UNFCCC's tradition of including all states in key drafting stages.[29] Once again, the legal-multilateral adoption in Cancun of the Copenhagen Accord's impulses suggested the combination of compelling extra-legal factors. These included the recognition by a broad range of parties, and the secretariat, of inherent value in the legalistic appearance of an agreement, and the powerful influences to which the construction of such an agreement opens. There was also the sense—among some parties and observers—that prior institutional principles and practices were deeply inconvenient, if not debilitating. These factors continued to play out in debates over "legal form" as the international climate regime evolved toward Paris and beyond.

Institutional Divisions and Shared Responsibility: The Question of "Legal Form"

The Cancun Agreements set a clear course toward "pledge and review" legality in the Convention's next major phase, but they did not entirely close debate over the character and reach of future commitments to reduce emissions, which had so derailed negotiations in Copenhagen. That debate continued to turn, in part, on the meanings of CBDR and equity: the very principles lauded by the Maldives and the United Nations Office of the High Commissioner of Human Rights (OHCHR) in

---

[27] See Gray 2010; Khor 2010a.

[28] Werksman 2010.

[29] Khor 2010a.

their approbation of the UNFCCC as a venue for rights-respecting, international, cooperative responses to climate change (Chap. 2). But economic and climatic developments had changed the playing field of negotiations. The KP Annexes, which had long served as a cornerstone for developing nations' and civil society groups' advocacy for just global climate governance came to seem increasingly illegitimate, while no new provisions emerged to take their place.

From the close of the Cancun COP to the opening of the following year's, in Durban, the question of "legal form"—the binding or non-binding status of mitigation commitments, and the possibility of retaining country groups reminiscent of Kyoto's Annexes—took center stage. US representatives would continue to play a leading role in efforts to shape the next phase of the Convention, insisting on what they called "legal parity" or "symmetry" among developing and developed country parties as a pre-condition of any new agreement with legal status.[30] They tended to frame these demands in pragmatic terms. As State Department Climate Envoy Todd Stern put it, "[o]ur goal is very simply to design a regime that is going to have the capability to actually help us solve the problem."[31]

The US and its allies in the "Umbrella Group" negotiating bloc[32] were by no means alone, however, in classing Kyoto's Annexes as a significant obstacle to effective regulation, and a potent stumbling block for international cooperation. Indeed, several climate justice advocates I interviewed, who remained staunchly critical of developed countries' strong-arm negotiating tactics, readily acknowledged problems with the Annexes. For one thing, scientific analyses had already shown that the developed country signatories to the convention could not, on their own, reduce emissions sufficiently to prevent the "dangerous anthropogenic interference with the climate system" against which the Convention's mandate orients it.[33] In addition, developed countries were beginning to abandon their Kyoto commitments. Arresting the progress of ongoing climate change would require broad based action, including by developing countries. (Incidentally, by multiple metrics developing countries had already made more ambitious pledges to cut emissions, in 2011, than their developed country counterparts.[34])

---

[30] Werksman 2011.

[31] Goldenberg 2010a.

[32] As the UNFCCC describes it, "The Umbrella Group is a loose coalition of non-EU developed countries which formed following the adoption of the Kyoto Protocol. Although there is no formal list, the Group is usually made up of Australia, Canada, Japan, New Zealand, Norway, the Russian Federation, Ukraine and the US." UNFCCC 2014.

[33] For example, Anderson and Bows 2011.

[34] Kartha and Erickson 2011.

Further, the crude division of parties into Annexes seems to have significantly contributed to a stalemate in the negotiations. It was hardly in the economic interest of countries protected under non-Annex I status to voluntarily "graduate" to Annex I, and extremely challenging politically for smaller developing countries to undermine the stance of the G77 + China negotiating bloc, whose positions have been shaped by its members' non-Annex I status under the KP.[35] As one analyst I interviewed put it at the time, indexing frequent invocation of the Protocol in rigid party and civil society positions—and explaining his own decision to step back from negotiations after over twenty years—"all the work I want to do has to come after the battle of Kyoto."

In the Annexes, then, the KP institutionalized a particularly rigid, two-tiered ontology of socio-spatial separation, with lasting implications for the meaning of CBDR and even, arguably, for the Convention's potential success. On the one hand, the Annexes protected developing countries from shouldering responsibility for climate change they did not cause, even as they faced its most deleterious impacts, leaving open the possibility of those countries' emissions *increasing* as they attained development thresholds in line with cosmopolitan norms and human rights standards. On the other hand, the Annexes made it more difficult for the Convention to adapt to changing socio-ecologies of international development and climate change, instead fixing relations of responsibility and capacity as they appeared in 1997. The contradiction between China's protected status under the Annex system and its emergence as the top yearly GHG emitter and the world's second largest economy was its most obvious anachronism. Moreover, the BASIC countries (Brazil, South Africa, India, and China) emerged after Copenhagen as a more formidable negotiating bloc, at times undercutting the influence of the wider collection of developing countries in the G77. As the BASIC bloc realized its diplomatic power, however, maintenance of the KP's development ontology appeared still more important to other developing nations and their civil society allies, since the vast majority of those countries classified as "developing" have attained nothing comparable to the BASIC countries' economic might, emissions, or independent influence.

The KP's "legal form" therefore remained compelling for many, despite its obvious liabilities. From perspectives prioritizing mitigation, the Protocol's binding

---

[35] Interviewees noted, for instance, that CBDR, as understood in the Kyoto Protocol's Annexes, is inseparable from the power of the G77 negotiating bloc at the UNFCCC. Indeed, the Least Developed Countries' alliance with the EU in Durban in 2011 represented the first consequential fracturing of the G77 bloc, considered the main factor in adoption there of the "Durban Platform." See below.

targets remained, though unmet and inadequate, a far better model than the emerging voluntary framework of Copenhagen and Cancun. For the UNFCCC's "Least Developed Countries" (LDCs) and civil society actors like many in development organizations, the KP's particular encoding of CBDR stood crucially for protection of the most vulnerable and least responsible: those whose interests would be elided even more readily in a regime structured by "legal symmetry" and nationally driven flexibility.

Of Convention Principles and Geopolitical Practice

As the example of Kyoto's Annexes illustrates, operationalizing the Convention principles of CBDR and equity in specific, legally constructed relations of responsibility was consequential both politically and ecologically. Because those principles were originally left undefined, they were rhetorically malleable, as countries debated the outlines of what would become the next legal phase of the international climate regime. Though parties attempted to leverage their different interpretations of equity and CBDR in divergent visions for the future allocation of duty and entitlement, it would be the US' position that ultimately took precedence, as the delegation navigated both international and domestic imperatives.

In 2011 a group of experts from the BASIC countries released a co-authored report on the meaning of the equity principle. The countries' environment ministers had argued jointly in Copenhagen that "a global goal for emission reductions should be preceded by the definition of a paradigm for equitable burden sharing." Though the countries' expert teams differed on some of the particulars of their formulae, they agreed that under the Convention equity should be conceptualized in terms of historical responsibility, "sustainable development," and per capita analysis of the benefits and burdens of anthropogenic climate change.[36] By relevant criteria, they argued, Annex I countries have overoccupied the available "atmospheric space" within which *all* nations must be afforded rights to emit GHGs in the course of development. The report represented an intensification of long-standing efforts to define CBDR and equity in relation to countries' historical contributions to climate change, which have included think tank Eco-equity's Greenhouse Development Rights framework and Bolivia's climate debt proposal, championed by climate justice advocates in civil society (see Chaps. 2 and 4 respectively).[37]

---

[36] Winkler et al. 2011.

[37] See also Friman 2007.

While acknowledging the capabilities that new and rising development status can afford, these arguments emphasize the lasting accumulation of emissions from long-industrialized countries, and therefore their culpability for climate change harm, as constituting a central differentiating element in responsibility.[38] Unsurprisingly, long-industrialized developed countries have instead emphasized current emissions and economic capabilities, particularly those of the BASIC countries. As parties debated a new agreement, developed countries also highlighted research showing that, in aggregate, developing countries had overtaken developed ones in yearly emissions, and are likely to contribute an increasing share. Thus, the US legislature's original rejection of mitigation obligations under the KP was based in part on the exclusion of China—which was already beginning its ascent in economic stature and emissions—from similarly binding commitments.[39]

Throughout subsequent negotiations, the US and its allies often avoided formal discussion of equity and CBDR, to the point of reportedly attempting to exclude their mention from agreements.[40] Stern's on-record comments on the topic in Copenhagen, however, seemed calculated to inflame advocates of historically differentiated responsibility: "we absolutely recognize our historic role in putting emissions in the atmosphere up there that are there now … but the sense of guilt or culpability or reparations, I just categorically reject that."[41] The US has sometimes seemed hostile toward the legalism of international negotiations in general, as well as the particular principles enshrined in the Convention and expressed in the Kyoto

---

[38] The capabilities/culpabilities distinction mirrors that between "response-able" and responsible parties, discussed in Chap. 1. Cf. Cameron and Bevins 2012. In practical terms, capable and culpable parties are in all but exceptional cases one and the same. CSO Equity Review 2018.

[39] It bears repeating, however, that the territorial assignment of emissions itself is dependent on specific political economic assumptions. Should gases released through industrial processes, for instance, be counted against the ledger of the country in which those processes occur, those whose citizens' consumption practices drive those industrial processes, or those where accumulation from production and consumption ultimately focus? See Chap. 1 and Bergmann 2013.

[40] Pickering et al. frame their analysis of "fairness in the next global climate agreement" around a quip journalists recorded from a huddle of negotiators during the final wordsmithing hours of Durban's COP, widely attributed to Todd Stern, that "if equity's in, we're out." See Pickering et al. 2012. In the end, neither equity of CBDR appeared in the "Durban Platform for Enhanced Action," the central document agreed at COP17, which set the direction for the next major phase of the treaty, though the principles are referenced in the Paris Agreement. Gray 2011.

[41] Broder 2009.

and Bali agreements. After Copenhagen, Stern opined that "one of the frustrations in dealing on the international level is that a lot of focus can be paid to debating whether a particular idea is consistent or not consistent with such-and-such article of a previous agreement; a lot of attention can be paid to proposals or positions that are not very well tethered to reality."[42] For Khor and others, the comment suggested that "the US does not want to negotiate within the UNFCCC with due respect for the legal tenets and provisions of this treaty."[43]

The US team would quickly pivot, though, finding ways to mobilize those same provisions explicitly, to support its own position. In 2011 Stern offered, "the way we look at it, [equity is] a concept that embodies fairness….From a US point of view, any agreement that's going to be legal has got to have legal symmetry, by which I don't mean that everybody has to do the same thing or that developing countries that are growing at a rate of 7 or 8 or 9 percent a year will have to make absolute reductions."[44] At COP18 in 2012 he went further, unexpectedly acknowledging that a future agreement will require parties to directly address CBDR and equity, and stating that "the United States would welcome such a discussion."[45] While the US held firm against Kyoto-style Annexes and negotiated targets in the next phase of the international regime, that is, it also signaled willingness to participate in a new agreement with some type of legal status, as demanded by its developing country interlocutors, and to some variant of the hallowed Convention principles. On the other hand, it was clear to all in Cancun and later that, for the US to join, those principles would have to be operationalized in terms of an all-inclusive "symmetry" between parties' obligations, and that "flexibility" and "self-differentiation" would necessarily apply to all. Accordingly, it was the US' preferred conception of equity and flexibility, in "Nationally Determined Contributions" (NDCs) for all nations, that became the core of the new regime in Paris.

If the legal form sketched and signed as the Paris Agreement directly reflects the principle of voluntarism and the form of "fairness" the US has long pursued in climate negotiations, those positions in turn reflect the fact that forceful international law is itself a highly problematic notion in the US context, for institutional and cultural reasons as well as economic ones. During a lengthy interview, a former negotiator offered candid clarification of how some on the US team understood and

---

[42] Goldenberg January 14, 2010a.
[43] Khor 2010b.
[44] Friedman 2012.
[45] See Roberts 2012.

experienced the impulses toward socio-spatial and governmental separation embodied in the international system and in arrangements like the Paris Agreements, which tend to insulate the US and other powerful emitters in particular:

> For the US, but I think also for China and other major countries ... the river flows domestic to international....
>
> Can you name a case where on anything of significant economic import, on a significant scale ... the way we do things in the US came about as a result of the international trend or international agreement or international foray? ... I can think of multiple where it worked the other way around ... We took action [on ozone depleting chemicals] nine years before the Montreal Protocol. Endangered Species Act preceded [the Convention on International Trade in Endangered Species]. The Marine Mammal Protection Act of 1972 preceded by 13 years the International Whaling Commission moratorium on whaling....
>
> I'm not saying we fixed them, because we haven't, but ... we took action as a result of our national dynamics. And in many of those cases, we carried that forth to an international venue and essentially said, "Will you please do what we're doing?"....
>
> The EU has a very different tradition. It works top-down ... they try to negotiate things at an international level ... [And] that's the whole set-up for the UNFCCC....
>
> [W]hy, year after year, do we continue to buy into this line from the EU, from the media, from the NGOs, that say "a treaty is so close ... if we could just get it right at the international level then we'll solve the problem?" ...It's not going to happen until we get our act together here.

Economic conditions, he argued, remained unripe, however, for US leadership on climate-changing pollution:

> You can come up with the other instances where you have industries ready to sell you their new products, make money off [new regulations].... In the case of ozone depleting substances, industry was largely aligned with [regulators] because they saw not only a lack of economic harm, but in many cases, they saw economic opportunities themselves in switching products....
>
> Carbon doesn't have to do with any of this stuff.... Because it's a global thing, a global problem, the premise is ... you're going to have some treaty. We tried it in '97 [with the Kyoto Protocol and] in The Hague, 2000, in the Netherlands at the end of the Clinton administration.... we're still trying to make it top-down.
>
> *Interviewer: We just don't have a history of that?*
>
> For any issue—environmental, pollutants—let alone the basis of our economy.[46]

On their own, these comments might be interpreted as instances of American exceptionalism, hubris, or even agency capture. But if the US delegation's position is

---

[46] Sunstein develops a similar argument explaining the differential successes of the Montreal Protocol and UNFCCC in terms of economic impacts. See Sunstein 2007.

explained there in logical terms, its actual role in climate negotiations is also effectively limited by the constitutional constraints placed upon its executive branch power by a similarly empowered, and often highly resistant, congress.[47] The Paris Agreement reflects these constraints not only in its reliance on flexible NDCs instead of legally binding negotiated targets, but in its formal status: under US law it is not a "treaty" *per se*, which would require ratification by congress, but an executive act taken independently. In the climate regime, therefore, political possibilities at the US national scale have proven constitutive of those at the international scale.

Despite the theoretical and pragmatic appeal of an inclusive, symmetrical regime based on what Stern described in 2013 as "norms and expectations as distinguished from rigid rules,"[48] however, the effects of institutionalizing these tendencies toward national priorities under the UNFCCC are worrisome. Shortly before the opening of the Cancun COP, the UN Environment Program (UNEP) released the "Gigatonne Gap" report, an extensive and now annual modeling study conducted by an international team of independent climate science centers, which evaluated the likely impact of the voluntary emissions reduction pledges gathered under the Copenhagen Accord.[49] The report, which was the subject of two overattended side events in Cancun, showed that even at the upper end of their potential those pledges fell far short of emissions reductions necessary to maintain safe levels of atmospheric greenhouse gases, as defined by the IPCC. The finding was unsurprising, given the lack of coordination with respect to a science-based target for total emission reductions under Copenhagen's "pledge and review" model, but the scale of the shortfall—five gigatons of atmospheric carbon equivalent in excess of "safe" levels—was sensational enough to become a prominent talking point for many organizations and parties.[50]

Emissions have continued to rise, and achieving the commitments thus far entered under the Paris Agreements (by both developed and developing countries) would only supply about a third of the reductions scientists deem necessary to keep global temperatures below the long-accepted benchmark of a two degree rise in the industrial era. That benchmark is itself considered by many observers to exceed

---

[47] Paterson 2011; MacNeil 2013.

[48] US Department of State 2013.

[49] UNEP 2010.

[50] Observation at COP16. See, for example, the several webpages of posts by the Climate Action Network tagged with "Gigatonne Gap" since 2010, CAN n.d.

livable thresholds for vulnerable communities. Reaching the two degree goal is thought to require reducing emissions, beginning no later than 2020, on a trajectory dramatically below that expected if countries' policies align with or even moderately improve upon their Paris Agreement NDCs. Achieving a target of 1.5 °C, as small island states, most in the climate justice community, and many scientists now demand, will of course require still greater action.[51] The Paris Agreement included provisions on adaptation and loss and damage as well, which would be supported at the international level primarily through climate finance mechanisms, but commitments have been decidedly low in this area as well, and implementation toward meeting them has lagged.[52]

Loving the KP: Legalism as Counter-Hegemony at the UNFCCC

As the international regime tacked back from the rupture in Copenhagen toward its erstwhile legal-multilateral tradition, and the US' uncompromising position on legal form became clearer, so did the stakes—for other negotiating parties and observers—of legality itself in global climate governance. In the run-up to COP17 in 2011, for instance, World Resources Institute's Jacob Werksman wrote[53]:

> A legally binding agreement is the highest form of expression of political will that the international community can bestow. Such agreements have been shown to bring a range of benefits through domestic ratification, the creation of international institutions, and the attention of high-level government officials, civil society, businesses, the media, and the broader public.
>
> Legally binding commitments within those agreements, such as specific targets and timetables, can create the confidence necessary to drive regulation, send clear market signals, and channel investments for a low-carbon future.

Legal status, that is, would express strong, shared approbation for, and recognize the moment of whatever plans and relations between countries a new agreement

---

[51] UNEP 2018; IPCC 2018.

[52] Climate finance is difficult to quantify, and even to clearly define, but estimates place the $100B commitment made in Copenhagen far below the amount necessary to address developing country infrastructure needs by 2020. World Economic Forum 2013. Over the 2015–2016 period, 81% of climate finance is estimated to have been raised in the same country in which it was spent, rather than transferred internationally, as many justice advocates demand. Climate Policy Initiative 2018.

[53] See Werksman 2011.

might enshrine, potentially helping to enable and mobilize the capacities of non-governmental and intergovernmental institutions in coordinated climate change responses. Legal commitments could mean more operationally, however, since without them, as an interviewee noted at the time, "there is no real impetus for people to honor those pledges."

But as the re-casting of equity and CBDR before the Paris Agreement suggested, the role of UNFCCC legalities is more complex, and deeply connected with both the content and the context of negotiations. On the one hand, it is clearly the details of binding commitments and their implementation, rather than their legal status, that determine their relationship to actual climate futures. As the interviewee quoted above made sure to add, "[y]ou have to remember that the problem we set out to solve was that we need to stabilize the global climate. Sometimes that's forgotten. Sometimes we think that we're here to reach agreements …. that's not the purpose." On the other hand, Convention principles play important structuring roles within the fray of negotiations, in part by providing developing countries with sanctioned channels for agonistic engagement on the unequal playing field of international relations.

Between sessions at COP15, I briefly engaged a G77 negotiator on principles and strategies for "justice" under the UNFCCC. Immediately producing a dog-eared copy of the Convention, she indicated the heavily underlined passage, from Article 3, on CBDR, suggesting that all anyone needed to know was "right here." Nodding, an experienced NGO staffer commented on the exchange later, in her Washington, DC office. "To talk to developing country negotiators you have to know your Convention inside and out, because they will walk you through the articles every single time; it's always back to the Convention, because that's what they have." She expanded, taking up the theme of CBDR's centrality for developing countries:

> It's not the most elegant of policy solutions, but … the Kyoto Protocol is the only international legally binding instrument. It has a very clear mandate and requirement for a second commitment period, and I think that there was a real concern that if you collapse those two tracks [agreed in Bali] you obviate some of the foundational principles that guide people's engagement in [climate negotiations]
> ….People were clinging to these legal principles that were contained – like Common but Differentiated, and second commitment periods – in the mandates, because it's the only tool that they have…. It's the one thing they can cling to, and they fight like hell for it, for years.

Particularly in the period between Kyoto and Paris, Convention principles like CBDR underpinned a form of subaltern legal mobilization at the UNFCCC[54]: in

---

[54] For subaltern legal mobilization as a strategic category, see Santos and Rodriguez-Garavito 2005. "Globalization from Below."

moments like the Bali COP, they aided developing countries in leveraging and focusing an effective counterweight to the power politics developing country negotiators and their governments can leverage.

At Copenhagen power politics ruled. Among the examples the DC NGO worker
offered was the removal of the very negotiator I'd spoken with in Copenhagen, a
prominent and respected diplomat, from her national delegation at the behest, reportedly, of the US team. "The kind of divide and rule pressure tactics that were
employed by developed countries on developing countries were extraordinary," she
continued. "On the floor of Copenhagen, on the very last night there were attempts
to remove trade preferences for developing countries if they didn't sign up to the
Accord." During COP15 and immediately afterward, such charges might have been
more easily dismissed. The release of WikiLeaks cables later confirmed them,
along with other pressure tactics operative beneath the procedural gloss of the negotiations.[55] The interviewee concluded:

> [I]n this particular forum ... thinking about legal and institutional questions and abil-
> ity to draw on those principles, I think developing countries are attempting to use the
> very few levers of power that they have – which are contained in the Convention – up
> against these phenomenal media and economic power houses.
>
>   Lumumba [Di-Aping, former Sudanese Ambassador], who chaired the G77, was
> not staffed at all .... G77 has no media support. A few of these countries have only
> two, or three, or four people on their delegations up against 300 from the US. Their
> ability to speak and communicate about their positions on the international stage is
> very, very limited.

Economic inequality between negotiating parties (and between civil society delegates) is a constant, consequential background condition at climate negotiations.
Late one evening during COP16, I shared a taxi with subdued delegates from two
Latin American governments, from the isolated luxury beachside conference center and hotel where the negotiations took place to the small downtown pension
where we were all staying. My room there had cost less than twenty-five US dollars
a night. Delegates from more well-resourced parties stayed in rooms in the conference complex, and worked or relaxed in private meeting facilities rented there.

As rules, structures, procedures, and sometimes principles, law between countries can provide a hedge against the exercise of power based in economic inequality. Such hedges can also wither and fail, as they arguably have in the re-allocation
of responsibility and the diminishment of ambition under the UNFCCC. The

---

[55] See Carrington 2010; Lister 2010.

outlines of what would become the Paris Agreement emerged, instead, from powerful developed countries mobilizing legality in the service of their own aims.

Negotiating Futures

Appeals to legality by developing countries and their civil society allies in the UNFCCC reflect a wider phenomenon in international politics and governance, theorized as subaltern legal mobilization.[56] Where protective legal rules do obtain, as they did under the Kyoto Protocol, a focus on legality can grow from and foster strong discursive links between institutional advocacy and broader mobilization efforts by stakeholders and allies. The convergent character of such strategies can inspire undue faith, however, in the potential of "law" to flatten uneven playing fields like that of international relations.[57] As reneging in Copenhagen morphed into renegotiating the legal meaning of key Convention principles, the role of law at the UNFCCC was clarified: not a bulwark, or an institutional and political weapon of the weak alone, but an ideologically loaded discursive and procedural terrain for struggle between contending parties motivated, in large measure, by nationally defined economic imperatives.

The successful push led by the US and its partners for consensus recognition of the Copenhagen Accord's impulses in Cancun suggested the felt importance of legal encoding in the UNFCCC. Paradoxically, demands for legality can be understood as helping fuel the US' counter-mobilization, which utilized the energy in those demands to formalize its own vision of responsibility in the Paris Accord's legal trappings, which could then be celebrated as a groundbreaking achievement for multilateral cooperation. Because agreements, principles, and targets with legal status under the Convention are constructed through participants' interactions in negotiations against a backdrop of deep inequalities, powerful participants can shape and re-shape relations of recognition and responsibility there, and seal them with the imprimatur of international legality. The US' and its allies' influence over the shape of a post-2020 global climate regime, and the likely impact of that regime on development and security elsewhere, illustrate what are ultimately co-constitutive ties between geo-economic power and international legal frameworks. The Copenhagen Accord initiated an apparent withdrawal from the rule of law. Later, in Cancun and Durban, the sovereign voluntarism and "symmetry" by which

---

[56] See Santos and Rodriguez-Garavito 2005.

[57] On relations between legal and social mobilization, see McCann 1994; Lobel 2007.

the Accord re-articulated equity and CBDR was transposed to a conventionally legalistic register, amidst the US' hardline negotiating posture, the imperative of legalistic legitimacy felt by other parties, and the institutional embrace of a pragmatic understanding of "consensus." The targets and principles introduced, and "noted," under the Copenhagen Accord substantially guided the development of the Paris Agreement and the emergence of a new regime with "legal force."

As the Paris Agreement serves to redefine relations of responsibility between those parties historically responsible for anthropogenic climate change and those most subject to its deleterious impacts, it will reconfigure the geopolitical landscape within the ambit of the UNFCCC and further, given the bargaining practices that shape negotiations, and the economic implications of mitigation, adaptation, loss and damage. It will present parties with new, contingent resources for interstate debates, as the founding principles of the Convention did for developing countries in Bali. The re-definition of relations through this co-constitutive legal-political process, however, suggests less promise for the protection of the most vulnerable. That is, by pushing the UNFCCC process outside of the field of the rule of law in Copenhagen, the US initiated a re-writing of the legal terrain for a future global political ecology and economy, through climate change governance. As it has taken shape since, that terrain is even more heavily marked by the socio-spatial and governmental separations of national sovereignty, which pose powerful limits on the pursuit of accountability, protection, or redress across borders.

The renegotiations that lead to the Paris Agreement also demonstrate the constitutive role domestic political and economic conditions play in possibilities for international climate policy, particularly where economically powerful countries are concerned.[58] That the US has so profoundly shaped the international climate regime from the start illustrates how multiple relationalities are intermingled in the institutional settings that have made climate justice and effectiveness so difficult to achieve. The commitment to Westphalian sovereignty embedded in international law and politics, which allows countries to act independently in relation to shared challenges like climate change, and the mirroring commitment to national interest within the US (and other countries) constitute socio-spatial separations with important governmental implications. They underwrite the insulation of domestic political actors from stakeholders abroad who are disproportionately affected by emissions and governmental decision making within nation-states. In this sense, the Trump government's withdrawal from the Paris Agreement that previous US negotiating teams shepherded to fruition did not constitute as noteworthy a change

---

[58] See Paterson 2011; MacNeil 2013.

of direction, as some observers have suggested, in the longer tradition of American power politics in climate governance.[59] Instead it continued that tradition, while raising the ante with characteristically more explicit forms of isolationism, disregard for legality, transactionalism in the international context, and political opportunism domestically.

On the one hand, Trump's distancing of the US from the international climate regime also raises the stakes in what had been recognized since COP15 as the crucially important politics of climate action and justice in other settings: across and between national and subnational polities and in civil society.[60] Activism and advocacy developed quickly and creatively in these settings, as coming chapters demonstrate, despite the noteworthy difficulties they entail.[61] First, to the extent national politics determine the possibilities of international action they demand new approaches of justice advocates, since national frames can obscure climate injustice, making it more difficult for advocates to render it legible and frame it in relation to established legal and policy paradigms. That is, climate-related inequities at the global scale and between nations may not even appear in within-country analyses of impact and responsibility. Global and international concerns, moreover, are typically difficult to address through domestic political institutions and debates. Those debates require advocates to frame interests differently, and engage via different formal modes and informal norms of participation than they do in other national contexts or the international forum provided by the UNFCCC. Climate justice advocacy and activism within countries, therefor, requires specific grounding in social, institutional, and political contexts, as Chaps. 4 and 5 examine in greater detail.

On the other hand, the UNFCCC remained an important target for climate justice seekers after Copenhagen's turn, in part because of the forum it provides for building alliances and supporting developing countries' interests, but also because of the felt sense, among many advocates and delegates, that global cooperation remains a necessity for effective climate change responses.[62] Through the period following the Paris Agreement and the Trump government's renunciation of it, the UNFCCC, country parties, civil society groups, and other actors have focused on principles and mechanisms by which the international climate regime might maintain legitimacy and increase effectiveness going forward.

---

[59] *The New York Times Magazine* 2018.
[60] Cf. Routledge 2011.
[61] Derman 2013, 2014, 2018.
[62] Derman 2014.

UN leaders and influential negotiating parties have underscored that the Paris Agreement continues with or without the US participation. Vulnerable developing countries and their civil society allies have lauded justice-related provisions in the agreement, including recognition of CBDR as a basic tenet, continuing discussions about responses to loss and damage, and acknowledgment of 1.5 degrees Celsius as the preferred range for global temperature increase. In response to the inadequacy of commitments and lack of legal levers, mechanisms have emerged to foster trust and cooperation among countries, and thereby increase ambition and improve outcomes. Under the convention proper, formal "ratcheting" provisions include periodic "stocktakes" during which parties will convene to evaluate pledges in relation to mitigation needs. A "Tanaloa dialogue" process brought countries' representatives together in parallel with negotiations to air concerns and build trust. Following Trump's disavowal of the Agreement in 2017, US mayors, governors, and citizens began attending COPs and other Convention meetings in greater numbers, with the explicit aim of supporting the institution and implementing US commitments despite federal inaction and efforts at sabotage. The Convention also faced formidable new challenges, however, including widespread political and economic crises that have rendered international coordination more difficult, as fraught debates over the meaning of a "just transition" away from fossil fuels during COP24, and repeated re-locations of COP25 during 2018 and 2019 illustrated.

Conclusion

A global climate regime with "legal force" but reliant on countries' voluntary "self-differentiation" and flexible implementation of mitigation plans represents the achievement of a nomospheric project: a project of world making and unmaking that draws upon law.[63] That project was accomplished by deepening the institutionalization of socio-spatial and governmental separations, which insulate polities and decision makers from affected people, enshrining global unaccountability in the name of national sovereignty. Narrating the achievement of the Paris Agreement illustrates the co-constitutive link between international law and geo-economic power in the context of climate governance, and the constraining influence of US political possibilities on international ones.

Even so, by encoding approbation, censure, and entitlement in universalizing moral terms, the language of law can also provide discursive frames for a politics of the disempowered with few parallels and wide resonance.[64]

---

[63] Delaney 2010.

[64] Goodale and Merry 2007; Delaney 2010; Derman 2014.

Coming chapters demonstrate that nomothetic political languages—those redolent of law[65]—therefore persist amidst and in spite of the absence of formal avenues to recognition or redress. Simultaneous with the re-negotiation of responsibility under the UNFCCC, a social movement for climate justice came of age on its fringes and in public political spaces beyond them.[66] That movement has drawn upon the language of law as well as other discourses in its connective analyses, principles, and practices to construct a global politics with growing resonance for groups around the world. The separations that beset the UNFCCC are socially constructed but materially consequential. The politics of connection is about addressing, undermining, and superseding those separations.

Bibliography

Anderson, K., & Bows, A. (2011). Beyond "Dangerous" Climate Change: Emission Scenarios for a New World. *Philosophical Transactions of the Royal Society A: Mathematical, Physical and Engineering Sciences, 369*(1934), 20–44.

Arts, B. (1998). *The Political Influence of Global NGOs: Case Studies on the Climate and Biodiversity Conventions*. Utrecht: International Books.

Bergmann, L. (2013). Bound by Chains of Carbon: Ecological–Economic Geographies of Globalization. *Annals of the Association of American Geographers, 103*(6), 1348–1370.

Betsill, M. M., & Corell, E. (2008). *Ngo Diplomacy: The Influence of Nongovernmental Organizations in International Environmental Negotiations*. Cambridge, MA: MIT Press.

Bodansky, D. (2010). The Copenhagen Climate Change Conference: A Postmortem. *American Journal of International Law, 104*(2), 230–240.

Bond, P. (2011). *The Politics of Climate Justice*. London: Verso and Pietermaritzburg: University of KwaZulu-Natal Press.

Broder, J. M. (2009, December 11). U.S. Climate Envoy's Good Cop, Bad Cop Roles. *The New York Times*. Retrieved from http://www.nytimes.com/2009/12/11/science/earth/11stern.html

C40 Cities Climate Leadership Group. (2015). *History of the C40*. Retrieved February 20, 2015, from http://www.c40.org/history

Cameron, E., & Bevins, W. (2012, December 14). *What Is Equity in the Context of Climate Negotiations?* World Resources Institute. Retrieved from http://www.wri.org/blog/2012/12/what-equity-context-climate-negotiations

---

[65] Delaney 2010.

[66] See, for example, Bond 2011.

CAN (Climate Action Network International). (n.d.). *Tag: Gigatonne Gap*. CAN International. Retrieved February 21, 2015, from http://www.climatenetwork.org/category/tags/gigatonne-gap

Carrington, D. (2010, December 3). *WikiLeaks Cables Reveal How US Manipulated Climate Accord*. Retrieved February 20, 2015, from http://www.theguardian.com/environment/2010/dec/03/wikileaks-us-manipulated-climate-accord

CIEL (Center for International Environmental Law). (2011). *Analysis of Human Rights Language in the Cancun Agreements (UNFCCC 16th Session of the Conference of the Parties)*.

Climate Policy Initiative. (2018). *Global Climate Finance: An Updated View 2018*. Climate Policy Initiative. Available at: https://climatepolicyinitiative.org/publication/global-climate-finance-an-updated-view-2018/

Cole, W. (2007, June 3). Australia to Launch Carbon Trading Scheme by 2012. *Reuters*. Sydney. Retrieved from http://www.reuters.com/article/2007/06/03/environment-australia-climate-dc-idUSSYD26700820070603

CSO Equity Review. (2018). *After Paris: Inequality, Fair Shares, and the Climate Emergency*. Manila/London/Cape Town/Washington, et al.: CSO Equity Review Coalition. [civilsocietyreview.org/report2018] https://doi.org/10.6084/m9.figshare.7637669

Delaney, D. (2010). *The Spatial, the Legal and the Pragmatics of World-Making: Nomospheric Investigations*. Abingdon: Routledge.

Derman, B. B. (2013). Contesting Climate Injustice During COP17. *South African Journal on Human Rights: Climate Change Justice: Narratives, Rights and the Poor, 29*(1), 170–179.

Derman, B. B. (2014). Climate Governance, Justice, and Transnational Civil Society. *Climate Policy, 14*(1), 23–41.

Derman, B. B. (2018). "Climate Change Is About US:" Fence-Line Communities, the NAACP and the Grounding of Climate Justice. In T. Jafry (Ed.), *Routledge Handbook of Climate Justice* (pp. 407–419). London: Routledge.

Fisher, D. R. (2010). COP-15 in Copenhagen: How the Merging of Movements Left Civil Society Out in the Cold. *Global Environmental Politics, 10*(2), 11–17.

Friedman, L. (2012, January 9). *NEGOTIATIONS: Can a New Structure Based on the Notion of "Equity" Replace the Kyoto Pact?* Retrieved February 21, 2015, from http://www.eenews.net/stories/1059958230

Friman, M. (2007). *Historical Responsibility in the UNFCCC*. Retrieved from http://www.diva-portal.org/smash/record.jsf?pid=diva2:233089

Galanter, M. (1974). Why the "Haves" Come Out Ahead: Speculations on the Limits of Legal Change. *Law & Society Review, 9*(1), 95–160.

Gardiner, S. (2011). Climate Justice. In J. S. Dryzek, R. B. Norgaard, & D. Schlosberg (Eds.), *The Oxford Handbook of Climate Change and Society* (pp. 309–322). Oxford: Oxford University Press.

Goldenberg, S. (2010a, January 14). *Next Few Weeks Vital for Copenhagen Accord, Says US Climate Change Envoy*. Retrieved November 6, 2014, from http://www.theguardian.com/environment/2010/jan/14/climate-change-us-envoy-copenhagen

Goldenberg, S. (2010b, November 30). *Cancún Climate Change Summit: America Plays Tough*. Retrieved February 20, 2015, from http://www.theguardian.com/environment/2010/nov/30/cancun-climate-change-summit-america

Goodale, M., & Merry, S. E. (2007). *The Practice of Human Rights: Tracking Law Between the Global and the Local.* Cambridge: Cambridge University Press.

Gore, A. (2006). *An Inconvenient Truth: The Planetary Emergency of Global Warming and What We Can Do About It.* Emmaus: Rodale.

Gray, L. (2010, December 12). *Cancun Climate Change Summit: Bolivians Dance to a Different Beat, but Fail to Derail the Talks.* Retrieved from http://www.telegraph.co.uk/earth/environment/climatechange/8197539/Cancun-climate-change-summit-Bolivians-dance-to-a-different-beat-but-fail-to-derail-the-talks.html

Gray, L. (2011, December 11). *Durban Climate Change: The Agreement Explained.* Retrieved from http://www.telegraph.co.uk/news/earth/environment/climatechange/8949099/Durban-climate-change-the-agreement-explained.html

Grubb, M. (2011). Durban: The Darkest Hour? *Climate Policy, 11*(6), 1269–1271. https://doi.org/10.1080/14693062.2011.628786.

Herbert, S., Derman, B., & Grobelski, T. (2013). The Regulation of Environmental Space. *Annual Review of Law and Social Science, 9*(1), 227–247.

(IPCC) Intergovernmental Panel on Climate Change. (2018). *Global Warming of 1.5° C.* Retrieved from https://www.ipcc.ch/sr15/

James, C. L. R. (2001). *The Black Jacobins: Toussaint L'Ouverture and the San Domingo Revolution.* London: Penguin UK.

Johl, A., & Duyck, S. (2012). Promoting Human Rights in the Future Climate Regime. *Ethics, Policy & Environment, 15*(3), 298–302.

Kartha, S., & Erickson, P. (2011). *Comparison of Annex 1 and Non-Annex 1 Pledges Under the Cancun Agreements.* Stockholm Environment Institute. Retrieved from http://www.oxfam.org.nz/resources/onlinereports/SEI-Comparison-of-pledges-Jun2011.pdf

Khor, M. (2010a). Complex Implications of the Cancun Climate Conference. *Economic and Political Weekly, XLV*(52), 10–15.

Khor, M. (2010b, February 12). *After Copenhagen, the Way Forward.* Retrieved February 21, 2015, from http://www.twn.my/title2/climate/info.service/2010/climate20100212.htm

Klinsky, S., & Winkler, H. (2014). Equity, Sustainable Development and Climate Policy. *Climate Policy, 14*(1), 1–7. https://doi.org/10.1080/14693062.2014.859352.

Kolbert, E. (2006). *Field Notes from a Catastrophe: Man, Nature, and Climate Change.* New York: Bloomsbury Pub.

Lister, T. (2010, December 10). *WikiLeaks: Cables Reveal Pessimism in Climate Change Talks.* Retrieved February 20, 2015, from http://www.cnn.com/2010/US/12/10/wikileaks.climate.change/

Lobel, O. (2007). The Paradox of Extra-Legal Activism: Critical Legal Consciousness and Transformative Politics. *Harvard Law Review, 120*, 937–988.

Lynas, M. (2007). *Six Degrees: Our Future on a Hotter Planet.* London: Fourth Estate.

MacNeil, R. (2013). Alternative Climate Policy Pathways in the US. *Climate Policy, 13*(2), 259–276. https://doi.org/10.1080/14693062.2012.714964.

Mayors Climate Protection Center. (2005). *U.S. Conference of Mayors Climate Protection Agreement.* Retrieved February 20, 2015, from http://www.usmayors.org/climateprotection/agreement.htm

McCann, M. W. (1994). *Rights at Work: Pay Equity Reform and the Politics of Legal Mobilization.* Chicago: University of Chicago Press.

Minow, M. (1991). *Making All the Difference: Inclusion, Exclusion, and American Law.* Ithaca: Cornell University Press.

Mitchell, T. (2011). *Carbon Democracy: Political Power in the Age of Oil*. London: Verso Books.

Paterson, M. (2011). Selling Carbon: From International Climate Regime to Global Carbon Market. In J. S. Dryzek, R. B. Norgaard, & D. Schlosberg (Eds.), *The Oxford Handbook of Climate Change and Society* (pp. 611–624). Oxford: Oxford University Press.

Patterson, J. (2011). *Nature's Fury—Chronicling the Devastating Effects of Climate Change in the US South*. May 8. Available at: https://climatejusticeinitiative.wordpress.com/2011/05/08/nature%E2%80%99s-fury%E2%80%94chronicling-the-devastating-effects-of-climate-change-in-the-us-south/

Pickering, J., Vanderheiden, S., & Miller, S. (2012). "If Equity's In, We're Out": Scope for Fairness in the Next Global Climate Agreement. *Ethics & International Affairs, 26*(04), 423–443.

Roberts, T. (2012, December 11). *Doha Climate Change Negotiations: Moving Beyond the Dueling Dinosaurs to Bring Together Equity and Ambition*. Retrieved February 21, 2015, from http://www.brookings.edu/blogs/up-front/posts/2012/12/11-doha-negotiations-roberts

Routledge, P. (2011). Translocal Climate Justice Solidarities. In J. S. Dryzek, R. B. Norgaard, & D. Schlosberg (Eds.), *The Oxford Handbook of Climate Change and Society* (pp. 384–398). Oxford: Oxford University Press.

Santos, B. de S., & Rodríguez-Garavito, C. A. (2005). *Law and Globalization from Below: Towards a Cosmopolitan Legality*. Cambridge: Cambridge University Press.

Schueneman, T. (2010, July 1). *Yvo de Boer Leaves UNFCCC Post "Appalled" by International Inaction*. Retrieved February 20, 2015, from http://globalwarmingisreal.com/2010/07/01/yvo-de-boer-leaves-unfccc-post-appalled-by-international-inaction/

Smith, N. (2003). *American Empire: Roosevelt's Geographer and the Prelude to Globalization*. Berkeley: University of California Press.

Sunstein, C. R. (2007). Of Montreal and Kyoto: A Tale of Two Protocols. *Harvard Environmental Law Review, 31*(1), 1–65.

UNEP (United Nations Environment Program). (2010). *The Emissions Gap Report 2010*. Retrieved February 21, 2015, from http://www.unep.org/publications/ebooks/emissionsgapreport/

UNEP (United Nations Environment Program). (2018). *Emissions Gap Report 2018*. Retrieved January 2019, from UN Environment website. http://www.unenvironment.org/resources/emissions-gap-report-2018

UNFCCC (United Nations Framework Convention on Climate Change). (1997). *The Kyoto Protocol*. Retrieved from http://unfccc.int/kyoto_protocol/items/2830.php

UNFCCC (United Nations Framework Convention on Climate Change). (2007). *Bali Road Map*. Retrieved February 20, 2015, from http://unfccc.int/key_documents/bali_road_map/items/6447.php

UNFCCC (United Nations Framework Convention on Climate Change). (2014). *Party Groupings*. Retrieved February 21, 2015, from http://unfccc.int/parties_and_observers/parties/negotiating_groups/items/2714.php

United Nations. (1992). *United Nations Framework Convention on Climate Change*. U.N. Symbol A/AC.237/18.

US Department of State. (2013, October 22). *The Shape of a New International Climate Agreement*. Retrieved February 21, 2015, from http://www.state.gov/e/oes/rls/remarks/2013/215720.htm

Vidal, J. (2010, February 10). *Yvo de Boer Steps Down as UN Climate Chief to Work for Accountants KPMG.* Retrieved February 20, 2015, from http://www.theguardian.com/environment/2010/feb/18/yvo-de-boer-climate-change

Walsh, B. (2010, December 10). Climate: Why the U.S. Is Bargaining So Hard at Cancún. *Time.*

Werksman, J. (2010, December 17). *Q&A: The Legal Character and Legitimacy of the Cancun Agreements.* Retrieved February 21, 2015, from http://www.wri.org/blog/2010/12/qa-legal-character-and-legitimacy-cancun-agreements

Werksman, J. (2011). *The Challenge of Legal Form at the Durban Climate Talks.* World Resources Institute. Available from: http://www.wri.org/blog/2011/11/challenge-legal-form-durban-climate-talks

Winkler, H., Jayaraman, T., Pan, J., de Oliveira, A. S., Zhang, Y., Sant, G., ... Raubenheimer, S. (2011). Equitable Access to Sustainable Development. In *Contribution to the Body of Scientific Knowledge: A Paper by Experts from BASIC Countries, BASIC Expert Group: Beijing, Brasilia, Cape Town and Mumbai.* Retrieved from http://www.mapsprogramme.org/wp-content/uploads/Basic_EASD_Experts_Paper.pdf

World Economic Forum. (2013). *The Ways and Means to Unlock Private Finance for Green Growth.* Available at http://www3.weforum.org/docs/WEF_GreenInvestment_Report_2013.pdf

Part II

Transnationalism and Grounding

Friday, 6 PM, Durban: It is the eve of the Global Day of Action, at the midpoint of COP17. Shadows lengthen as a motley handful of activists and observers descend the hill from the University of KwaZulu-Natal (UKNZ), the main site of open gatherings during this COP. We have been invited to the evening meeting of the Rural Women's Assembly (RWA), with its 650 plus delegates from 9 countries.[1]

The visitors are late, however, and the meeting has ended. Only a few organizers remain. Clustered at tables beneath a cavernous tent at one side of a parking field, they put finishing touches on signs for tomorrow's march. Members drift from the tent, toward a folding table where boxed dinners are being handed out.

From the other direction come tense voices: organizers of another group that has brought several hundred members to Durban, to participate in a conference at the University and demonstrate en masse outside the COP. A grave-faced young organizer yanks boxes from an overstuffed hatchback, as his comrades depart in the direction of another large tent, at the far end of the lot. Disappointed to have missed the Women's discussion and doubtful of

---

[1] Hargreaves 2012.

what else this night might reveal about the experience of these groups during the COP, I succumb to the urge to be useful, approach the heavily loaded car, and am put to work.

Soon the boxes are piled behind a row of plastic chairs at one end of the second tent. The organizer opens one and exclaims in frustration. Its contents—t-shirts freshly printed for the group to wear in tomorrow's march—have not been packaged by size. Another helper is quickly hailed to help unpack, sort, and stack the shirts on every available surface, as the tent fills noisily. Before taking their own evening meal, the group's members have had to file past that of the RWA outside. They take seats, many clearly eager to finish the business portion of the evening and get on with supper. The organizers have much they wish to accomplish, however, and the food is not here yet.

Representatives from a handful of transnational civil society groups with accreditation to enter the COP arrive and address the group: after updates and critiques on the negotiations, they suggest messages the group might deliver during tomorrow's march. Then organizers review procedure for how members will stay together and safe along the route: captains have been designated, each a contact point responsible for a contingent of marchers. Next, they debrief about a smaller action from earlier today: "What did we learn? How did members of the group feel?" Responses vary, from inspired to frustrated to unrelated.

One of the transnational activists takes the mike again. "Is it OK if I lead us in a very popular song that we sing around here … Is it OK?" he booms, receiving scattered assent. Then his powerful voice floats over the assembly, inviting the welcome release of shared tension and fatigue. The energy in the tent is trebled, as whistles, catcalls, and voices rise in a familiar call and response:

My mother
—My mother was a kitchen girl
My father
—My father was a garden boy
That's why
—That's why I'm a socialist
I'm a socialist, I'm a socialist…

Later, meals arrive and are spread on a plastic table. Members of the group queue eagerly. Each collects a shirt with their dinner.

As the group disperses, the leader of the t-shirt brigade offers a ride back to the University, urging me to wait while he and others gather in a third, smaller tent for a final, brief planning session. It is late now, and the group is small. Everyone looks tired. Several express anger at the lack of cooperation from other groups participating in tomorrow's mass mobilization. Largely, though, conversation centers on how to transport the group assembled here to the starting place of the march and to ensure its needs for rest and food are better met than they have been today. Another silent observer waits just outside the circle of beleaguered organizers.

Saturday: Some 8000 people join in the march, which stops before a large conference center, the COP location, in a formidable security detail. Members of The Rural Women's Assembly deliver a memo to the UNFCCC's Executive Secretary and the South African Environment Minister, amidst a throng of cameras.

Along the route, a phalanx of marchers clad in the jersey of the host country's COP staff—members of the ruling party's youth wing—attempt to delay and separate the second group that gathered in the parking field from the rest of the crowd. The group's members manage to complete the route, though one of its leaders is physically assaulted.

Three nights later the other silent attendee from the pre-march strategy session turns up at an informal presentation of the BASIC experts group's equity paper, in a well-appointed private venue downtown. Over the plentiful hors d'oeuvres he offers that, though sympathetic, he was in the tent through personal rather than vocational ties.

Although material conditions and political opportunities for stakeholder groups like the RWA differ strikingly from those of credentialed delegates and scientists, each is bound to the other in social relations, places, and the whorl of an evolving climate. Some, like the transnational activists who visited with marchers before the Day of Action, are at pains to make those ties more apparent and meaningful within international debates.

Regional constituencies like the RWA and others, and transnational groups like those who addressed them before the march in Durban are central actors in the construction of a popular politics of connection, which has largely taken shape outside and on the fringes of the COPs and other intergovernmental meetings. Their roles are different but each essential in attempts to bring the voices of affected people into global governance. Their participation is also crucial to the emergence and dissemination of connective analyses: those that render climate justice a broad political stake, by linking climate change with the compromising of livelihoods and life chances for marginalized people around the globe, and with the social relations and processes that produce and reproduce disproportionate vulnerability.

The next two chapters focus on specific strategies, principles, tensions, and achievements in the politics of connection as it has developed beyond international institutions. Transnational spaces and logics of climate justice expose and challenge the separating tendencies that have compromised international political and legal frameworks but they also lay bare profound dilemmas inherent in amplifying the voices of disproportionately affected groups and inspiring broad-based mobilization. Those dilemmas—and activists' efforts to resolve them—suggest that a truly connective politics must recognize and leverage existing geographically and historically specific political identities and infrastructures, as well as the revealed and emergent bonds of the Anthropocene.

On the Outside

4

Morning, the first day of COP16 in Cancún. Murals grace concrete walls downtown, miles from the secured Moon Palace complex where negotiations are beginning: "Protege el medio ambiente"; a cartoon of the Earth cradled in human hands; "Bienvenido COP16 – Al Estado de Quintana Roo donde la Anarquía y la Corrupción va siempre hacia adelante."[1] Here, in a compact municipal park, La Via Campesina and a handful of Mexican and transnational groups will host farmers and activists from across the continent and the globe during the two-week-long negotiations.

Banners and flags are already hung in advance of the caravans. Several frame movement slogans or demands: "Continual Growth is the Ideology of Cancer"; "Peasant Agriculture Cools the Earth"; "No GMO corn – Food Sovereignty!". Others proclaim the names of groups involved in mounting this public forum: La Via Campesina; Friends of the Earth International; Oilwatch; the Mexican National Liberation Movement; the Assembly of Affected People. At least one banner demands the release of political prisoners.

A massive tent looms behind the band shell. Several small sleeping shelters are pitched beneath it already, together with an extensive outdoor

---

[1] "Protect the environment" and "Welcome COP16 – The state of Quintana Roo where Anarchy and Corruption will always go forward."

© The Author(s) 2020
B. B. Derman, *Struggles for Climate Justice*,
https://doi.org/10.1007/978-3-030-27965-3_4

kitchen and a line of porta-potties. Under the band shell as well, infrastructure for the gathering is being raised. A makeshift radio tower and a trio of plywood and plexiglass booths to one side of the sea of folding chairs will broadcast live translations of testimonial and analysis delivered from the stage in Spanish, English, Italian, Portuguese, and Greek.

As the morning wears on, organizers welcome new arrivals to the park with wrist bands. Many exchange warm greetings. A sense of anticipation builds. Soon, cameras of every size emerge amidst a general drift toward the band shell.

Before the stage, observers crowd around as women and men in woven cloth, braids, and the green bandanas of La Via Campesina move deliberately around an open area on the ground. Slowly, their work takes shape: within a circle of oranges, a cruciform of homespun; atop it, an arrangement of gourds, baskets full of grain and seed, ears of corn, clay chalices of oil and earth, a pile of squash, a conch; tall candles at every corner. Finally, with the lighting of this Mystica and a brief dedication, the "Global Forum for Life, Environmental and Social Justice" begins.[2]

Beyond secured negotiating rooms a simultaneous set of open forums and demonstrations, like those held in Cancún, highlight a different set of participants and discourses in the broader conversation about climate change. Less rule-bound and more visibly diverse than intergovernmental meetings, these "outside" events, as many participants call them, are nevertheless structured quite intentionally. Such gatherings are also frequently generative, playing host, for instance, to formative networking during COP6 in The Hague, the founding of the Climate Justice Now! (CJN!) network in Bali, and widening awareness of social movement energy and critiques in Copenhagen and Paris.[3]

This chapter examines how "outside" spaces, together with the ideas and practices that inform them, challenge the governmental, socio-ecological, and sociospatial relations that have comprised and compromised many mainstream responses to climate change. The first section analyzes settings like Cancún's Global

---

[2] See summary and final declaration of the summit at La Via Campesina 2010a.

[3] See also Chatterton et al. 2013, on the outside mobilizations of Copenhagen.

Forum as spaces of connection: accounting for organizers' aims and means in convening and structuring such gatherings, especially as they center on stakeholder inclusion, visibility, and multi-directional exchanges of knowledge.[4] Social and spatial strategies in which "outside" forums and mass demonstrations play a key role are primary concerns for many advocates and activists of climate justice, as their comments from meetings and interviews attest. Examples of events and interactions "on the outside," grounded in observation as well as participant interviews and published accounts, encapsulate some of the wider dynamics of transnational mobilization.[5]

The second section turns to specific analytical themes that have served as touchstones for the politics of connection in open forums and demonstrations, and sometimes inside intergovernmental meetings. Grounded in texts and campaigns by advocates, activists, and lay participants in "outside" venues, these themes exemplify connective analyses within and across relational domains. Together, these two sections illustrate how transnational spaces and analyses of connection have helped both to articulate critique and generate alternatives: by socializing, politicizing, and historicizing the extensive ties and uneven geographies of climate change.

The final section of the chapter centers on a series of challenges organizers face in convening constituents and forging shared positions through connective analyses. These hurdles include the material needs of regional groups gathered during international meetings; the alignment of many voices, with each other and the terms of intergovernmental debate; the potential for cooption by elites; and the resonance of movement principles in relation to the geographically and historically specific circumstances of far-flung stakeholders.

Spaces of Connection

"Outside" gatherings and the practices that structure them further a variety of interrelated political aims. Participants in these "civil society" or "peoples" spaces have included large numbers of un-credentialed local and regionally based stakeholders to climate governance debates, many of them members of existing movements and/

---

[4] Cf. Routledge 2017; Routledge and Cumbers 2009 on "convergence spaces."
[5] This chapter emphasizes examples from Durban, during COP17, where conditions of time and access to those with local knowledge enabled particularly fruitful observation and analysis of what are wider dynamics in transnational climate justice mobilization.

or groups on the frontline of impacts, as well as activists and community representatives from other parts of the world. Organizers work to assemble such diverse constituencies during intergovernmental meetings, and create opportunities for dialogue, alliance-building, and influence. As one explained in Durban, during a discussion devoted to strategy:

> Our CJ [climate justice] movement is not separate from other justice movements out there. That CJ movement is an economic justice movement, is a social justice movement, is a political justice movement. Climate brings all of our movements together, and we are building and developing. That's why we have these civil society spaces.... Here, the job of our movements is to build, connect, and make relationships.
>
> The conversations, the meetings that happen are about building from the ground up, whether it's local campaigns against extractives, or looking at transformations in food and agriculture, and in energy. It's looking at what the solutions are.... Our policies and positions come from real positions from our communities: from real experiences.... We formulate those, and we bring them into the UNFCCC space.

Just as bringing the positions of diverse outsider groups into the exclusive settings of governance requires engaging them in dialogue and constructing shared statements, indexing the legitimacy of those statements is thought to demand mass displays of stakeholder presence. In an interview, a staff member of a transnational organization described the role of broad-based participation within her group's wider strategy during intergovernmental meetings:

> [T]he outside piece is incredibly important... not only at the climate talks themselves, but also in generating political pressures in key countries at home. There's a couple of functions that mobilizations can serve. One is a demonstration of political interest and enthusiasm: sharing information and galvanizing interest and participation. That's an important goal in and of itself. The second, of course, is to exercise some political pressure on key political targets.

Local marches during COP meetings and simultaneous actions by allied groups elsewhere are both key actions, therefore, which organizers facilitate by coordinating "outside" gatherings as well as Global Days of Action (Fig. 4.1). As another organizer quipped in Durban, "[w]e are going to win if we can build a global movement," evidenced, he estimated, in the visible presence of 500 million people, marching simultaneously in cities around the world.

Assembling social movement groups, members of affected communities, and broader publics for open forums and actions demands tactics for enabling mobility and claiming space.[6] Two such tactics employed by climate justice organizers are

---

[6] Cf. Routledge 2017.

Fig. 4.1 Protestors prepare for the Day of Action during COP17 in Durban, South Africa, in December 2011. Simple slogans like "leave the coal in the hole" challenge the very under-pinnings of the global, fossil-fueled economy

caravans and camps. In preparation for the peoples' forum and mobilizations in Cancún during COP16, for instance, La Via Campesina, the Mexican Movimiento Liberacion National, the Mexican Electricians Union, and the Assembly of Affected Peoples organized caravans from six separate regions of the country. Participants traveled by bus along the caravan routes, stopping in cities, towns, and villages along the way to gather local accounts of environmental change and impacts from corporate and state-sponsored development projects. Their findings and personal testimonials made up a significant portion of the proceedings in the park. Many caravaners and their allies camped beneath the tent behind the main stage while they attended the Forum. Volunteers from the camp and the larger gathering shared cooking and cleaning duties in support of these who slept in the park during the COP.[7]

---

[7] See La Via Campesina 2010b.

The following year the Pan-African Climate Justice Alliance (PACJA) led one of the largest mobilization efforts to target COP17. The "Trans-African Climate Caravan of Hope" spanned some 7000 kilometers and ten countries, collecting travelers and messages and raising awareness in population centers along its path. Many participants traveled only a portion of the route, so that significantly more than the 300 activists, members of civil society, scientists, farmers, and journalists who ultimately arrived in Durban via the caravan took part.[8] A PACJA spokesperson described organizers' goals for the pilgrimage during a meeting at UKZN:

> The climate justice movement cannot survive if we don't involve farmers, if we don't involve women, if we don't involve those who are impacted heavily by climate change. Those are the pastoralists, and people who are at the forefront … of this problem and that was the spirit of this caravan.

In bringing individuals and accounts from "the forefront," PACJA sought to push for an international agreement, as the speaker put it, more "responsive to African realities." Accordingly, thousands along the route added their signatures to a petition demanding that COP delegates[9]:

1) Keep Africa and the world safe and prevent catastrophic climate change. Exert pressure on developed countries and ensure that they sign up to legally binding commitments that reduce emissions and limit global warming to well below 1.5 °C.
2) Share the effort of curbing climate change fairly. Demand domestic emission reductions by developed countries that are commensurate with science and equity, and enable a just transition in all countries.
3) Ensure polluters not the poor must pay. Developed countries must honor their obligations and pay at least 1.5% of their GNP to help the poor adapt and develop cleanly and sustainably.

Caravaners stayed to present and participate in discussions at UKZN and the Global Day of Action march. "Outside" forums and demonstrations in Durban were also importantly shaped by the presence of large contingents of the regionally-

---

[8] Gersmann and Vidal 2011; Mary Robinson Foundation—Climate Justice 2011; Kodili 2011.
[9] Quoted in Oloo 2011, together with a description of the deliberative process from which the Caravaners' demands emerged aligned.

based Rural Women's Alliance (RWA), the South African Democratic Left Front (DLF), and other visiting constituencies who stayed in dorms and elsewhere in the city during the meetings. Smaller groups and individual delegates from many more national, regional, and international organizations also lodged in dorms on the university campus, attending meetings there and shuttling to others downtown, including, for the few with access to it, the COP itself.

Social practices are just as important as spatial ones, like caravans and camps, on the "outside." The broad range of participants in these settings creates emergent opportunities for networking, enabling the linking of interests, campaigns, and constituencies, but many dialogues are both planned and intentionally structured. In addition to the reports and testimonials of caravaners and other participating groups, the extensive programming in open venues during COPs and other intergovernmental meetings have included accessible summaries, critiques, and analyses of negotiations and specific policy instruments, and discussions about alternatives as well as experiences and strategies of resistance.

As these examples suggest, interactions in "outside" venues emphasize the sharing of knowledge.[10] This includes knowledge of activity on the "inside" of negotiations, as well as experiences "on the ground" in local communities affected by climate change and related public or private initiatives. The multiplicity suggests another important facet of "outside" forums: climate change, its drivers, and its consequences are understood in these spaces through lay experience (such as the testimony of non-expert caravaners), and social, critical, and integrative perspectives as much as through scientific analysis and prediction. This epistemological openness facilitates participants' recognition of and mobilization around connections between those social relations and material conditions that structure life in communities, and those wrapped up in the production, impact, and governance of climate change.

In conjunction with presentations and mediated discussion, outside gatherings have also incorporated or even relied on emergent, horizontal interaction and principles of direct democracy to facilitate decision making and information sharing. This was the emphasis in the assemblies, vigils, and spontaneous meetings that took place throughout Durban's climate summit under the banner of Occupy COP17. Participants used a park opposite the Convention Center where negotiations took place, to which organizers had secured municipally sanctioned rights of use and given the name "Speakers' Corner." Occupy COP17 aimed to offer "a

---

[10] In this respect, these spaces fit the concept of Global Justice Network "convergence spaces" developed by Routledge and Cumbers 2009.

forum for those who wish to discuss and implement real and equitable solutions to climate change, with climate justice at the heart… open to all, operating on the principles of inclusiveness, openness, non-hierarchical organising and consensus decision-making."[11] Accordingly, gatherings in the park proceeded through social practices popularized by Occupy protests in Wall Street's Zuccotti Park and other locations around the world. The Speakers' Corner provided a visible space of dialogue and protest during the COP, and a meeting ground for a wide range of groups during the negotiations. Those who participated in assemblies and held their own gatherings in the park included members of 350.org and UNFCCC delegations from the Small Island Developing States group, as well as South and southern African civil society organizations such as the RWA and DLF.[12]

Prominent actions involving observers inside the COP also took shape that year through participatory practices associated with the Occupy movement. On December 9, 2011, as negotiations moved into their final stages, observers and a handful of party delegates blocked the main hallway of the Durban conference center and moved gradually toward a plenary, which was then in session. The demonstrators sang South African protest songs and echoed chants used earlier that week in the Global Day of Action.[13] When stopped by security guards, the group proceeded to debate its next move using the "human mic" technique for unamplified large group communication common in Occupy General Assemblies. Many chose to surrender their accreditation for the remainder of the meeting as they were escorted from the building; others remained and were carried out by security, apparently to be permanently barred from COPs. On the same day, observer Anjali Appadurai ended her statement to the COP plenary on behalf of "global youth" calling for the "mic check," which signals a speaker wishes to be heard during Occupy gatherings. Observers at the back of the room responded, using the "human mic" again to repeat and amplify in unison her calls for climate action and "equity now."[14]

Aims and practices associated with the "outside," then, are focused around fostering forms of connection: organizers strive to link the experiences and demands of affected people with the governmental capacities of official institutions, and

---

[11] Occupy Cop17 2014.

[12] Direct observation. The Speakers' Corner was also the center of an impromptu march on the day before the Global Day of Action. The website for the gathering also provided video statements by international activists. Occupy Cop17 2014.

[13] Direct observation. See Takver 2011.

[14] Direct observation. See Goodman 2011.

bring the issues and membership of diverse stakeholder groups together in broader understanding and alliance. They deploy spatial strategies of convening and presence, as well as interactional practices that facilitate participation, knowledge exchange, and crosscutting analyses. Among the effects of these connective moves is the partial breaking down of distinctions between "inside" and "outside," and this is indeed the intent of many of those involved. The organizer first quoted in this section, for instance, a member of an accredited but critical organization, argued, "those people who say that there are inside and outside spaces – I'm sorry but I think that is a sterile, and a wrong characterization. What we have is we have one struggle. We have one movement and in different spaces we use different tactics and different strategies."

Many of the groups involved in organizing open gatherings around COPs are accredited observers inside, active in official side events and press conferences, and the brief interventions afforded to civil society within negotiation sessions. Focused on the linking of inside and outside climate political spaces, or blurring the distinctions between them, these groups pursue a broader strategy aimed at prying open the doors of institutional power, enacting the liberatory potential embodied in the Janus-like figure of civil society.[15] In this connective role, they perform a notably different function, during COPs and other "environmentally" oriented intergovernmental meetings, than similarly critical groups have with respect to international forums like the World Trade Organization (WTO). As subsequent sections illustrate, although such linking strategies are reflectively considered, they nevertheless remain a subject of debate among critical observers of global climate governance.

"Outside," "open," "civil society," or "peoples'" forums have also produced collectively authored texts. Together with other seminal statements and analyses, these peoples' agreements have given impetus to continued mobilization and theory-building. The more influential of these texts and campaigns unfold connections that underpin growing solidarities, sharpen critique, and inspire alternative visions.

Connective Analyses and Campaigns

Major articulations of transnational climate justice politics include the founding statements of the Durban Group for Climate Justice and CJN!, the Bolivian proposal on climate debt and subsequent mobilization of that concept by civil society

---

[15] Cf. Swyngedouw 2005; see Chap. 6.

organizations, the Peoples' Accord and Declaration of the Rights of Mother Earth from the World People's Conference on Climate Change and the Rights of Mother Earth (WPCCC), and statements from key groups during COPs and other events such as Rio+20, the 2012 UN Conference on Sustainable Development. These statements combine biogeochemical and physical accounts of anthropogenic climate change with political economy and ecology, colonial and environmental history, feminist thought, and traditional knowledge. Those social and socio-ecological perspectives re-frame the (re)production of climate change and its disproportionate impacts, enabling visions of alternative futures that draw upon Indigenous, anti-racist, feminist, and class-based as well as ecological politics and principles. The focus on power that characterizes those longer-standing social movements inflects climate justice statements and campaigns, often in ways that reflect the times, places, and constituencies involved in their making. Amidst the reconstruction of climate governance through intergovernmental meetings, several of these campaigns and statements also invoke democratic norms that highlight stakeholders' rights, and their official representatives' responsibilities. Based on an eclectic set of principles, then, these relationally oriented analyses challenge institutional, social, and epistemological exclusions that have obscured the constitutive ties of climate injustice and subverted more inclusive, collectively driven, and ecologically grounded responses.

Public Power, Rights, and Frontline Voices

Many civil society groups have demanded increased accountability and capacity from public institutions in global climate governance, together with explicit limits on corporate influence. As Rio + 20 meetings opened, for instance, a broad civil society coalition launched "Dismantle Corporate Power and End the Impunity,"[16] a campaign analyzing UN bodies and their global meetings as deal-making sites for transnational capital. At approximately the same time, Friends of the Earth International (FOEI) released "Reclaim the UN," with a supporting petition endorsed by over 400 civil society groups.[17] The report argued that "[t]he UN is the most democratic and appropriate global institution" to address current global economic and environmental crises, but that it is also increasingly influenced by corporations and business lobbies, demonstrating that dynamic through six thematic

---

[16] Dismantle Corporate Power 2012.
[17] FOEI 2012.

case studies spanning multiple UN agencies and issue areas.[18] FOEI and its co-signatories called for a renewed commitment by the UN and member states to serve the public interest; recommended provisions to limit participation by business representatives, increase that of civil society, and improve transparency; and demanded a new legal framework to "hold companies accountable to environmental, human rights and labor rights law."[19]

Numerous open letters have also called for public power, like one from a coalition of youth civil society delegates to the UNFCCC Executive Secretary during the 2012 COP, which attacked the influence of fossil fuel interests over Canada and other parties' negotiating positions. The letter called attention to recent constraints on civil society activity in Convention meetings, juxtaposing it with norms governing countries' participation. It argued[20]:

> States like this are blocking progress in the name of an industry with the potential to break the planet.... Observer organizations can be penalized, and even removed from the convention if we violate the protocols for participation. Perhaps there should be a similar process for observers and parties whose mandates fundamentally contradict the convention.

Such efforts represent the widening uptake of watchdogging organizations' long-standing concerns with accountability and private influence.[21] They also resonate with increasing calls to democratize climate governance from critical observers and "outside" activists[22] like those who announced the Occupy COP17 forum, arguing that "[t]he very same people responsible for the global financial crisis are poised to seize control of our atmosphere, land, forests, mountains and waterways."[23]

Rights language is another prominent feature in transnational climate justice advocacy and activism. As Chap. 2 showed, rights-based arguments for mitigation, finance, and other forms of social and environmental protection in the face of climate change necessarily challenge territorially delimited, state-centric understandings of human rights duties, by taking greater account of global economic and ecological ties. Mary Robinson's rights-based analysis of climate justice, for

---

[18] Ibid, 4.

[19] Ibid, 7.

[20] Canadian Youth Delegation, joined by 137 delegates and 21 observer organizations, 2012.

[21] For example, Banktrack 2011; Global Witness 2011.

[22] See, for example, Vidal 2012.

[23] Occupy COP17 2011.

instance, begins with recognition of global socio-spatial and ecological connections: "[h]uman rights law is relevant," she writes, "because climate change causes human rights violations," and because "human rights make clear that government obligations do not stop at their own borders."[24] Nor are states' duties to respect, protect, and fulfill rights in the climate change context ahistorical for Robinson, who has argued that "the world that has benefitted from a carbon development [is] ... causing great suffering."[25]

A brief on "Human Rights and Climate Justice," released by a coalition of civil society groups,[26] built upon arguments for extra-territorial legal duties like those of Robinson and the Maldives (see Chap. 2), injecting them with more developed critiques of transnational capital and market-based governance. The brief argues that established human rights conventions already mandate that "[p]rivate companies must be subjected to climate- and rights-protecting regulations"[27] by signatory states. "Carbon markets – existing and proposed – should be re-examined from a human rights-based perspective," it asserts, adding, "[p]roposals to sell to the most polluting corporations rights over the soil of the poorest farmer and enable these corporations to continue polluting, are unacceptable."[28] The brief combines references to the Universal Declaration of Human Rights with the concept of climate debt, and quotes prominently from the WPCCC People's Agreement (see below).

Many in the wider transnational climate justice community have mobilized, both "inside" and "outside," to protect the rights of Indigenous peoples and other local communities against threats connected with forest-based carbon sequestration programs. Such programs are encompassed under the rubrics REDD and REDD+ in connection with UNFCCC provisions supporting both public and privately administered funding aimed at reducing emissions from deforestation and forest degradation in developing countries, and later additions promoting conservation and "enhancement" of "forest stocks," and "sustainable forest management."[29]

---

[24] International Council on Human Rights Policy 2009, iii–iv.

[25] Robinson addressing the crowd assembled at the destination of the Day of Action march in Copenhagen. Recording on file with the author. See Introduction.

[26] ActionAid et al. 2010b (Climate justice brief #12: Human Rights and Climate Change), released for the Cancún COP, with contributions from seventeen organizations including ActionAid, Friends of the Earth International, JS – Asia/Pacific Movement on Debt and Development (JSAPMDD), Jubilee South, PACJA, and the Third World Network.

[27] Ibid, 1.

[28] Ibid, 2.

[29] See, for example, UNFCCC 2019; REDD-Monitor 2019.

Beginning early in the development of those initiatives, legal advocacy groups called for assurances of free, prior, and informed consent, and other provisions enshrined in the Declaration of the Rights of Indigenous Peoples. Wider concerns and outright rejection of the programs, including by the "No REDD" campaign around which a wide coalition of civil society groups have rallied (see below), are tied to fears that such land-based governance initiatives risk deepening the ongoing dispossession Indigenous groups face, as they struggle to retain traditional territories and practices and contend with climate change-related impacts they have done little to cause.

As these REDD-related mobilizations and the PACJA organizer quoted above suggest, the issues, and to some extent voices, of directly affected communities lie at the center of many transnational climate justice efforts. This is perhaps most consistently the case with respect to Indigenous groups. As scholar-activist Tadzio Meuller notes[30]: "[m]ovements for climate justice ... rely strongly on the agenda-setting, the political leadership of often indigenous 'frontline communities' (that is, of those groups who are most directly affected by climate change as well as by the so-called 'false solutions' like emissions trading or agrofuels)."

Indigenous and allied activists have argued that, despite the diversity among Indigenous cultures and experiences, shared histories of subjugation and marginalization exacerbated by disproportionate impacts from climate change and responses warrant shared mechanisms of protection as well as solidarity. The Indigenous Environmental Network (IEN) and other coalitions are, accordingly, prominent within wider networks like CJN!,[31] as well as national and regional alliances for climate justice. Indigenous peoples' rights form central planks within coalitional analyses and platforms like the "Climate Justice Brief" series, the founding statements of CJN! and the Environmental Justice and Climate Change Initiative, Global Justice Ecology Project's influential formulation of climate justice, and the WPCCC texts. Even more crucially, perhaps, principles of Indigenous thought underpin some of the most widely traveled alternative visions associated with climate justice, as discussed below.[32]

The priority of frontline communities' experiences in transnational climate justice politics also has situational dimensions. One example is the strong focus on

---

[30] Mueller 2012, 72.

[31] See, e.g., IEN Executive Director Tom Goldtooth's statement quoted in the Preface.

[32] See Environmental Justice Climate Change Initiative 2002; Climate Justice Now! 2007; Climate Justice Now! 2010; Environmental Justice Leadership Forum on Climate Change 2015; Global Justice Ecology Project 2015.

local, national, regional, and continental perspectives that developed around the COP in Durban, South Africa, in 2011, the UN International Year for People of African Descent.[33]

Studies suggest that Africans as a group are among the least responsible and most affected by climate change.[34] Accordingly, many organizations participated in an overarching "Stand with Africa" campaign and worked to amplify efforts by African groups as part of COP17-related advocacy.[35] Transnational non-governmental organization (NGO) ActionAid participated in and publicized the PACJA caravan in a series of blog posts, for instance, while also releasing its own lengthy report, which critiqued the development of soil carbon markets as a "false solution" to the deepening precarity of African farmers.[36] ActionAid, Grassroots Global Justice Alliance, and other transnational networks also carried the stories of RWA members in their media feeds and disseminated the group's memorandum to the COP,[37] which articulated what Hargreaves called the "triple burden of race, class and geographical marginalization" borne by African women under climate change[38]:

> Rural women across Southern Africa are already reporting 20 per cent decreases in food production, and current trends tell us that if we fail to take action now, by 2020 we will have seen a 50 per cent loss in crop yields in our region. We produce 80 per cent of the food consumed by households in Africa. In the absence of support for us, we believe that local and national food security will be deeply threatened.[39]

Participants in Occupy COP17 issued a statement on the International Day of Action for Human Rights linking the geographical concentration of climate impacts in Africa with future global effects: "Our planet is changing, and with it the story of human rights. Here in Africa, the river beds are already drying and the seedlings that we watered for the future are wilting... Climate change is the tyranny of the present over the rights of the future. In Africa, the future is already here."[40] Later, Occupiers joined CJN! in calling the Durban COP's outcome a

---

[33] See United Nations 2011.

[34] See, for example, Heinrich Böll Foundation 2011.

[35] See Stand With Africa 2012.

[36] Kodili 2011; ActionAid 2011.

[37] Bote 2011; Grassroots Global Justice Alliance 2011.

[38] Hargreaves 2012, 9.

[39] Rural Women's Assembly 2011.

[40] Occupy COP17 2014.

"crime against humanity," by which "the richest nations have cynically created a new regime of climate apartheid."[41]

That the vulnerability of frontline communities is to a significantly degree socially produced—through longstanding relations of colonial and racialized oppression, in these examples—highlights the importance of socio-ecological analyses in understanding climatic injustice. Accordingly, these form an essential complement to governmentally connective campaigns for public control of institutions and a more thoroughgoing application of recognized rights.

Socio-ecological Principles

Transnational climate justice politics has drawn upon, combined, and developed socio-ecological principles from multiple sources, including "Third Wordlist," peasant, feminist, and labor movements; ecologically imbued critiques of capitalism; and traditional Indigenous thought.

The concept of climate debt underpins the calls of several civil society groups and countries in the Global South for industrialized nations and global elites to redress climate injustices through international financial transfers and progressive taxes. These justice advocates consider climate debt a component within a larger complex of ecological, social, and economic debts owed by "developed countries and corporations…to the poor majority."[42] Formal calculations of climate debt developed out of earlier analyses of unequal ecological exchange and ecological debt, which included the global flow of resources and distribution of ecological damage, and arrived at a reversal of financial lenders' accounts of debt owed by countries in the Global South to those in the North. The research collective Environmental Justice Organizations, Liabilities and Trade (EJOLT) traces the concept of ecological debt to early 1990s work by Latin American NGOs including Accion Ecologica, and subsequent development by FOEI (both also climate justice thought leaders). In international terms, ecological debt can be defined as[43]:

> (1) the ecological damage caused over time by a country in other countries or to ecosystems beyond national jurisdiction through its production and consumption patterns [and] (2) the exploitation or use of ecosystems (and its goods and services) over time by a country at the expense of the equitable rights to these ecosystems by other countries.

---

[41] Ibid, Climate Justice Now! 2011.
[42] ActionAid et al. 2010a. See also Roberts and Parks 2009; Third World Network 2009.
[43] EJOLT 2013.

While "[e]cological debt usually designates a public debt a country has towards other countries (foreign debt)," EJOLT adds that the concept "can also be used to calculate a debt (or liability) from a company (private debt) or a debt a nation has towards future generations (generational debt)."[44]

As formulated in an April 2009 proposal submitted by Bolivia to the UNFCCC, climate debt can be calculated in terms of historical and continuing emissions (occupying a share of "atmospheric space" available for the accumulation of greenhouse gases) together with the costs of adaptation in other communities made necessary by those emissions.[45] As ecological debt did for foreign debt ledgers, climate debt reframes longstanding debates within international negotiations over historical responsibility, casting adaptation funding as debt payment. The Bolivian proposal argues, therefore, "[d]eveloping countries are not seeking economic handouts to solve a problem we did not cause. What we call for is full payment of the debt owed to us by developed countries for threatening the integrity of the Earth's climate system"[46] by occupying more than their fair share of atmospheric space. The G77 developing country negotiating block took up the concept later in 2009.[47]

Climate debt also became a cornerstone in the analyses and proposals of several major civil society groups. Jubilee South – Asia/Pacific Movement on Debt and Development (JSAPMDD), for example, anchored its advocacy in debates over the sources and administration of international adaptation funding and technology transfer in its understanding of climate debt, arguing similarly that "climate finance is not aid or assistance but part of reparations that is long overdue."[48] One of the CJN! network's five core demands (see below) is for "massive financial transfers from North to South, based on the repayment of climate debts and subject to

---

[44] Ibid.

[45] Government of Bolivia 2009. "Developed countries and corporations owe a two-fold climate debt to the poor majority: For their historical and continuing excessive emissions – denying developing countries their fair share of atmospheric space – they have an 'emissions debt'; For their contribution to the adverse effects of climate change – requiring developing countries to adapt to rising climate impacts and damage – they have an 'adaptation debt.' The sum of these debts constitutes their climate debt, which is part of a larger ecological, social and economic debt owed by the rich industrialized world to the poor majority." ActionAid et al. 2010a. Historical responsibility being for many parties a central element of Common but Differentiated Responsibility and Respective Capabilities (CBDR-RC). See Chap. 3. See also Friman 2007.

[46] Government of Bolivia 2009, 47.

[47] Roberts and Parks 2009.

[48] Jubilee South 2010, 2.

democratic control." Civil society groups have also proposed and modeled mechanisms through which global elites could be held accountable for their accumulating climate debts, for instance by taxing the luxury emissions of individuals and service providers. The World Development Movement (later Global Justice Action) and ActionAid, for example, framed their advocacy for international taxes on financial transactions, shipping, and aviation to leverage climate finance for developing countries in terms of the necessity of repaying historical and mounting climate debts.[49]

In April 2010 some 30,000 representatives of social movements, NGOs, Indigenous communities, and governments gathered outside Cochabamba at the invitation of then Bolivian President Evo Morales, to address what his government and other organizers deemed the catastrophic outcomes of UN negotiations the previous year in Copenhagen. The WPCCC[50] produced two major texts: the *Peoples' Accord* and the *Proposed Declaration for the Rights of Mother Earth*, which articulated a set of principles and propositions that have also been widely invoked in transnational advocacy and mobilization for climate justice.

The first article of the *Declaration* lays out the ontological underpinnings of the texts' ecologically infused moral propositions: "Mother Earth is a unique, indivisible, self-regulating community of interrelated beings that sustains, contains and reproduces all beings... Each being is defined by its relationships as an integral part of Mother Earth."[51] Correspondingly, the *Accord* asserts, "[i]n an interdependent system in which human beings are only one component, it is not possible to recognize rights only to the human part without provoking an imbalance in the system as a whole. To guarantee human rights and to restore harmony with nature, it is necessary to effectively recognize and apply the rights of Mother Earth."[52] The *Declaration* therefore proposes "Mother Earth and all beings of which she is composed have...the right to maintain its identity and integrity as a distinct, self-regulating and interrelated being," and "[e]ach being has the right to a place and to play its role in Mother Earth for her harmonious functioning."[53]

---

[49] See, for example, World Development Movement and Jubilee Debt Campaign 2009; ActionAid 2009. ActionAid cites supporting analysis by Stamp Out Poverty: Stamp Out Poverty 2009; Stamp Out Poverty & Institute for Development Studies 2011. See also the Robin Hood Tax organization's analysis of climate debt: Robin Hood Tax 2010.

[50] See descriptions of the Cochabamba meetings in Building Bridges Collective 2010; Mueller 2012.

[51] WPCCC 2010b, 1.

[52] WPCCC 2010a, 2.

[53] WPCCC 2010b, 2.

The *Accord* locates the ultimate causes of the failure of negotiations and the seemingly inexorable progression of climate change in the capitalist global economy and the national governments that enable it. Capitalism, it asserts, seeks to sunder the fundamental nature-society ties captured in the concept of Mother Earth, and obscures the socio-ecological processes of extraction and exploitation that give rise to a changing climate along with wider social and ecological devastation[54]:

> The corporations and governments of the so-called "developed" countries, in complicity with a segment of the scientific community, have led us to discuss climate change as a problem limited to the rise in temperature without questioning the cause, which is the capitalist system.
>
> ….The capitalist system has imposed on us a logic of competition, progress and limitless growth. This regime of production and consumption seeks profit without limits, separating human beings from nature and imposing a logic of domination upon nature, transforming everything into commodities: water, earth, the human genome, ancestral cultures, biodiversity, justice, ethics, the rights of peoples, and life itself.
>
> Under capitalism, Mother Earth is converted into a source of raw materials, and human beings into consumers and a means of production….

The Cochabamba texts were disseminated widely, along with accounts of the gathering itself, inspiring and structuring later advocacy and mobilization efforts. In addition to critique of international negotiations and governance, the Cochabamba authors elaborated and fused pre-existing principles that movement leaders in both "inside" and "outside" spaces were able to hold up as alternatives or mechanisms of redress in the face of social disruptions tied to climate change, inappropriate governmental responses, and neglect.

The *Accord*, *Declaration*, and other initiatives that originated in the WPCCC were quickly brought into official intergovernmental forums through UNFCCC and other UN proceedings lead by Bolivia and its allies.[55] The Cochabamba gathering and its products also became touchstones for civil society forums, demonstrations, and the institutional engagements of key inside-outside linking organizations. La Via Campesina built on WPCCC principles to frame what would become the central public gathering during COP 16 in Cancún.[56] There, the People's Conference and

---

[54] WPCCC 2010a, 1.

[55] A committee presented the two texts to the July 2010 UNFCCC intercessional meeting in Bonn. Mueller 2012. Other initiatives launched in Cochabamba included proposals for an international environmental court and a global referendum on the rights of Mother Earth.

[56] La Via Campesina's invitation to gather in Cancún is reprinted as an appendix (noted as page xii) in Building Bridges Collective 2010.

the rights of Mother Earth (or "rights of nature" as some groups and speakers preferred) were ubiquitous evocations in meetings and mobilizations. For instance, activists in the park literally sung the praises of Cochabamba, changing the lyrics of the popular song "La Bamba" to describe the conference, and using the name of the city as the repeated refrain of the song's chorus. Marchers carried a banner with an image of the Earth from space and the one word "Cochabamba," and chanted the call "Obama, Obama, respeta Cochabamba!" Prior to and during the 2011 climate negotiations, rights of nature advocates gathered signatures on a worldwide petition, and planned and participated in numerous events both within and outside the COP.[57]

Speakers at the open forums during COP17 presented the rights of nature as an alternative to then-emerging "green" forms of capitalism inflecting UN discussions of mitigation, forestry, and the wider construction of what some finance and government leaders called a "green economy."[58] Former Bolivian ambassador to the UN Pablo Solón cast the rights of nature as a basis for opposition and, ultimately, transformation[59]:

> We need to come out of Durban with a clear statement … from different organizations – all of us, we that are here… that we propose a new system that recognizes this new relation with nature: this recognition that also nature has rights.
>
> If we don't develop this alternative we are going to be facing a defensive fight in six months in Rio de Janeiro [at the UN Conference on Sustainable Development]. So from our point of view, it's not only a theoretical discussion, it's a political discussion: What are going to be the main proposals on the table?
>
> We all say another world is possible. This new world will have to have a new kind of relation with nature if we are going to build a different society…. [W]hat began in Copenhagen, highlighting that we must change the system not the climate, has to be developed even more, to concrete alternatives of a new society: of a new relation between humans, and humans and nature

For Solón and others, the use of rights language in the proposal offered a direct challenge to the logic and institutional underpinnings of market-based responses to climate change. As another civil society network leader put it, "[I]n the negotiations … there's a WTO of the sky concerning privatization of air, privatization of trees, privatization of the soil…. It's about property rights…. That's why we've been embracing this new global coalition on the Rights of Nature, the Rights of Mother Earth: so that the Earth has standing, so that the Earth has jurisprudence."

---

[57] Global Alliance for the Rights of Nature 2015.

[58] See UNCSD 2011; Büscher and Arsel 2012.

[59] See announcement for and video documentation of Solón's talk at UKZN. Centre for Civil Society n.d.-a, b.

Mixed reaction to these calls to mobilize using the rights of nature illustrated some of the challenges inherent in the construction of transnational stakeholder coalitions (see below). Rights of nature provided a positive frame for widespread opposition to market-based climate governance initiatives, however. These too can be understood in terms of socio-ecological relational analysis. The exchange of offsets and permits to emit under these programs is founded on the assertion of equivalence among diverse mechanisms of greenhouse gas sequestration and release around the world. These equivalences obscure the social relations within which those mechanisms are embedded, in calculations of their aliquot contributions to the global atmospheric stock of climate change-producing compounds.[60] Markets and offsets, that is, can separate emissions from the contextual conditions that make legible the uneven allocation of benefit, risk, and harm with which they are intertwined. As in the racial and socio-economic inequities of polluting industrial location decisions documented by environmental justice analyses, abstracting greenhouse gas emissions and sinks from their social context can lead to disproportionate social distributions of benefits and harms. Moreover, in practice the aggregate mitigation potential of markets can be compromised by "leakage" associated with incomplete market coverage and the mobility of polluting industries, a product of the independent, spatially delimited administrative apparatuses upon which most markets depend, and by the influence of major emitters over consequential aspects of market design. Market regimes, critics warn, can thereby entail significant compromises of both effectiveness and justice.[61]

Other socio-ecological principles animate the development of alternatives in transnational climate justice debates as well. While some are more explicitly connected with anti-market analyses, economic concepts like climate debt, or ethical ones like the rights of nature, others foreground alternative ways of life, values, and livelihoods for human stakeholders.

The WPCCC *Peoples Agreement*, for instance, also "propose[s] to the peoples of the world the recovery, revalorization, and strengthening of the knowledge, wisdom, and ancestral practices of Indigenous peoples, which are affirmed in the thought and practices of 'Living Well.'" Here the phrase "Living Well" translates

---

[60] Larry Lohmann, Patrick Bond, and Michael Dorsey's work is among the most developed critical scholarship on this topic. See, for example, Lohmann 2008; Bond and Dorsey 2010; Bond 2011.

[61] Cf. Paterson 2011; Spash 2011.

the concept of *Buen Vivir*, elaborated and used by Latin American social movements to reject dominant conceptions of "development," and incorporated in national constitutions including those of Ecuador and Bolivia (along with the rights of nature).[62] *Buen Vivir* represents a multi-faceted description of the good life, based in community and underpinned by a non-binary, relationally oriented ontology, which focuses on interlinkages between nature and society, pointing away from consumption-based and widely networked economies. In this sense, *Buen Vivir* resonates with some eco-socialist conceptions of climate justice, as well as arguments tied to diverse philosophical traditions, which deploy notions such as "intrinsic value" and "relational value" to reorient economic and environmental stewardship in more ecologically caring, less anthropocentric ways while nevertheless still affirming the affective and material dimensions of human wellbeing as these are linked with non-human nature.[63]

A final strand of socio-ecological connectivity, exemplified by two campaigns, focuses on the livelihood concerns of large constituencies, both urban and rural, as they affect and are affected by climate change and the necessity of transition to new energy pathways. At UKZN and in the streets of Durban a coalition of trade unions, social movements, and environmental justice groups promoted the launch of a campaign for "One Million Climate Jobs" based on analyses of South African economic conditions and climatic vulnerabilities. Adapting a similar research-based initiative in the UK, the South African platform called for "a just transition to a low carbon economy to combat unemployment and climate change," by creating jobs that[64]:

(1) reduce the amount of greenhouse gasses we emit, to make sure that we prevent catastrophic climate change;
(2) build our capacity to adapt to the impacts of climate change (for example, jobs that improve our food security); [and]
(3) provide and secure vital services, especially water, energy and sanitation (this includes reducing wasteful over-consumption).

In a second example, La Via Campesina has argued for climate justice and effectiveness through rurally based labor and social structure. During COPs the

---

[62] WPCCC 2010b, 1. On the relationship between *Buen Vivir* and other political-ontological traditions, see Gudynas 2011.

[63] See Kovel 2007; Chan et al. 2016; Gudynas 2017.

[64] One Million Climate Jobs 2011.

group has linked its wider campaign for food sovereignty with the mitigation potential of small farmer agriculture, with slogans like "peasant agriculture cools the Earth,"[65] and "industrial agriculture heats up the planet, farmers are cooling it down."[66]

Together with the social and spatial strategies of open venues and demonstrations, these connective socio-ecological principles have underpinned the emergence of wider and deeper global solidarities, developed in the name of climate justice.

Networks of Global Solidarity

Networks linking multiple, geographically dispersed organizations have long been one of the staple structures supporting the work of civil society and social movement alliances in relation to global governance, in the environmental context and more broadly.[67] Like large NGOs with sufficient capacity to participate independently, broader networks composed of smaller development, faith-based, and environmental groups, many of them longstanding, are a prominent feature of civil society involvement in the climate convention, where they pursue a variety of aims and strategies. Networks share information and aggregate the input of far-flung allied stakeholders, many of whom are unable to send representatives to global meetings.[68] To the degree networks are able to achieve consensus among dispersed members, network representatives give those members voice when they participate in elite, official, and otherwise remote or exclusive forums.

Amidst broadening recognition of the disproportionate, uneven geographies of climate impact and responsibility, and rising urgency to quell emissions and address their consequences in the early 2000s, existing socio-ecologically oriented networks like La Via Campesina and FOEI began to prioritize climate justice within their broader set of concerns. Around the same time, networks

---

[65] Author's observation. See La Via Campesina 2011a.
[66] Author's observation. See La Via Campesina 2011b.
[67] See, for example, Keck and Sikkink 1998, 1999.
[68] Routledge and Cumbers 2009; cf. Derman 2014.

focused on climate justice at multiple scales began to emerge. CJN! has been one of the most important of these relatively recent networks in the international and transnational contexts delimited by forums like the UNFCCC and the simultaneous civil society summits discussed above. CJN! formed from twenty or more existing organizations (some networks themselves) during the Bali COP of 2007. Its founding statement describes its origins and ongoing commitment to analysis and presence both "inside" and "outside," as well as its core platform[69]:

> Peoples from social organizations and movements from across the globe brought the fight for social, ecological and gender justice into the negotiating rooms and onto the streets during the UN climate summit in Bali. Inside and outside the convention centre, activists demanded alternative policies and practices that protect livelihoods and the environment.
>
> In dozens of side events, reports, impromptu protests and press conferences, the false solutions to climate change – such as carbon offsetting, carbon trading for forests, agrofuels, trade liberalization and privatization pushed by governments, financial institutions and multinational corporations – have been exposed.
>
> Affected communities, Indigenous peoples, women and peasant farmers called for real solutions to the climate crisis, solutions which have failed to capture the attention of political leaders. These genuine solutions include:
>
> - Reduced consumption.
> - Huge financial transfers from North to South based on historical responsibility and ecological debt for adaptation and mitigation costs paid for by redirecting military budgets, innovative taxes, and debt cancelation.
> - Leaving fossil fuels in the ground and investing in appropriate energy efficiency and safe, clean, and community-led renewable energy.
> - Rights-based resource conservation that enforces Indigenous land rights and promotes peoples' sovereignty over energy, forests, land, and water.
> - Sustainable family farming and peoples' food sovereignty.
>
> Compared to the outcomes of the official negotiations, the major success of Bali is the momentum that has been built towards creating a diverse, global movement for climate justice. We will take our struggle forward not just in the talks, but on the ground and in the streets – Climate Justice Now!

---

[69] Climate Justice Now! 2007.

In its platform and calls to unity, the statement suggests some of the processes and relations involved in producing and deepening a world-encompassing, class-like division between socio-ecologically marginalized and privileged groups, as well as the necessity of developing solidarities both within and across that division. CJN! and allied networks have been crucial in hailing and mobilizing such a broad and diverse coalition to advance climate justice causes in institutional and wider political advocacy. In 2008 the network sent out a call for wider membership; by 2010 it included over 700 organizations and coalitions around the globe.[70]

Politicizing Global Climate Governance

Noting the ascendance of managerial, scientific, and consensual discourses and modes of participation since the 1990s, Swyngedouw and others have character-ized climate change governance as exemplifying a "post-political" condition.[71] And yet, as Chatterton et al. suggest, this perspective itself privileges the framings and practices of official processes such as those of the UNFCCC.[72] Empirically, transnational climate justice activists (and a few states) have articulated overtly critical, collective positions on climate change and coordinated responses to it, of-ten based in the open "outside" spaces of alternative summits and demonstrations. The relationship between such settings and the "inside" milieu of intergovernmen-tal fora is, as the evidence offered above suggests, of central concern to many activ-ists and observers.

"Outside" spaces have provided a locus for convergence among the "distinct trajectories"—to borrow Massey's term[73]—traveled by a diverse set of affected groups and publics, serving as a social base from which organizers have sought to challenge existing global climate governance by showcasing the presence and am-plifying the voices of historically marginalized constituencies in particular. Central features of organizing in outside venues therefore include the spatial and social strategies of open meetings, street demonstrations, caravans, and camps, each de-signed to facilitate inclusive deliberation and visibility.[74]

---

[70] Climate Justice Now! 2008, 2010.

[71] Swyngedouw 2007, 2010.

[72] Chatterton et al. 2013.

[73] Massey 2005.

[74] Cf. Routledge and Cumbers 2009 on similar gatherings as "convergence spaces," and Routledge 2017 on spatial strategies of activists, including the Paris COP of 2015.

A set of critiques and proposals animates and sometimes emanates from inter-action on the "outside," often crossing from or to critical civil society and state in-terventions within intergovernmental fora. These analyses challenge the political, epistemological, and ontological legitimacy of established climate governance, by confronting it with evidence of limits to growth; colonial, imperial, and capitalist dispossession; and the embeddedness of economy and society within a global, more-than-human relational matrix. Extending beyond critique of "business as usual," these analyses also proffer visions of more equitable alternatives that may better reflect the global socio-ecological ties that produce and reproduce anthropo-genic climate change and its disproportionate impacts. Together, these connective spaces, analyses, and campaigns radically expand the range of actors, perspectives, and principles at play in climate politics.

At the same time, the construction and mobilization of transnational solidarities for climate justice are not without their own difficulties and tensions. These include challenges inherent in involving and representing marginalized groups, risks of cooptation involved in the conditional embrace of official power, and pitfalls in seeking to assemble diverse coalitions around socio-ecological principles, the meanings of which are necessarily shaped by historical and geographic context.

Tensions in Transnational Climate Justice Advocacy

The afternoon session of a day-long conference program is beginning at the main "outside" meeting venue during COP17, and much of the crowd has not returned from the lunch break. The conference schedule is full, though, and the organizers can wait no longer. They close the doors and the keynote speaker, a US-based scholar-activist, begins.

Minutes later, singing can be heard outside: faint at first, now rising and becoming clearer. Then voices explode into the hall, accompanied by rhythmic clapping, in which much of the already-assembled audience joins. It is a contingent from the RWA. They continue to sing as they file in and take seats. The speaker resumes, welcoming the new arrivals and appearing en-ergized, if surprised, by their entrance. Two similar interruptions occur. As more local and regional constituents arrive, the organizers and speaker grow less sanguine in receiving latecomers.

Later, when presentations conclude for the day, the now large audience takes part in a wide-ranging discussion, by turns appreciative, inquisitive, and challenging. Questions are posed and statements delivered on theory and strategy as well as national political conditions, local agricultural practices, possible health effects of climate change, linguistic barriers, and the relative absence of African voices "outside" as well as in.

Singing interruptions during the key note lecture operated as assertions of presence and identity on the part of visiting constituent groups, while also challenging the tight structuring and one-to-many format of the conference. Audience contributions in the discussion period, which included direct challenges to panelists and organizers, linked the presence of visiting groups outside the COP with socioecological and historical conditions relevant in their own lives, sometimes more so than the topic of presentations or the climate negotiation itself.

Issues of positionality, representation, and power are bound up, in these outside settings, with the process of arriving at shared analyses, principles, and positions for organizers and constituents to take to the streets and linking groups to "the inside," in white papers and official statements. Those messages must resonate with the situated lived experiences of people affected by climate change—and those next in line—if they are to facilitate broad mobilization and guarantee the legitimacy of representative civil society groups. The discourse of global governance, by contrast, is often arcane, requiring something akin to translation to enable clear connection with social movement and community issues and demands.[75] International forums are also, by design, not grounded in geographical specificities: COPs are literally itinerant, touching down in cities around the world for a period of weeks. In contrast, local and regional presence on "the outside" is geographically and historically rooted, and in novel ways with every iteration of the globe-trotting negotiations. Finally, lay constituencies are in effect "crashers" to the COP party: organizers, not governments, must often secure space and resources for their safe and visible presence. These tensions each pose challenges for the construction of active, transnational climate justice solidarities, of which outside and linking organizers are in general well aware, but that nevertheless demand their continual attention.

---

[75] Cf. Hart 2013 on "translation" as a crucial political modality.

One Voice

To affect the trajectory of global governance by mobilizing a mass of stakeholders too numerous and ubiquitous to ignore, organizers must articulate a clear, unified position that can resonate among a diverse constituency drawn from widely distributed communities and movements. In seeking to do so, they contend with philosophical rifts, power differentials, and distinct political and socio-ecological concerns among local and national groups, as well as the complexities of governance, which are often ill-suited to summary or consensus-building among broad coalitions.

The UNFCCC, for instance, combines specialized language with complex structures and processes. As one activist complained, "negotiations have become this complicated maze, and a lot of people say they don't understand about this, and they don't get involved." Leaders seeking to gain influence for non-accredited stakeholders therefore work to render negotiations in accessible terms, and shape popular interventions so as to address issues within negotiations. Another member of an inside-outside linking NGO spoke about this work in an interview:

> I think the extent to which those mobilizations are useful to influence the policy discussions is how well tailored they are to the key leverage points in the negotiations, and that's always the hardest piece to figure out: in a way to communicate it that's effective in an outside mobilization, that speaks with an authoritative voice that brings on the concerns of the people on the ground. Because the people on the ground, the people that are marching and mobilizing, are the people that stand to be impacted the most: first and the hardest. The positions that are brought by those communities are, rightly so, the strongest, most radical demands. And I think the question is often how you translate that into something that policy makers can understand and work with.

Indeed, movement messaging has typically focused at a macro level, as in demands for "people over profits," keeping fossil fuels underground, prompt climate action, and justice. Some of the challenges of translation and coalition building were illustrated by mobilizations against REDD and provisions that would recast it as REDD+, which became a major focus both within COP16 negotiations and amongst demonstrators outside.

As the same interviewee put it, "[t]he REDD piece is extremely contentious for a lot of reasons." Many in the climate justice community identify dire risks and harms to Indigenous and forest communities' livelihoods, land claims, culture, and self-determination through REDD, as well as dubious assumptions of mitigation by avoiding expected emissions, and the potential for unequal exchange and speculative investment through offsets. On the other hand, heavily forested developing countries—which many justice advocates would like to support in negotiations—as well as some communities and local NGOs have seen potential in the programs for much-needed economic development through preserved, or even farmed, forests. REDD/REDD+ has therefore been the subject of sometimes heated debate: a wedge among otherwise potentially aligned constituencies, suggesting a wider set of challenges posed by differences in positionality that can be both material and ideological in origin.

"The opposition to REDD+ was a big deal, in large part because of the focus on carbon offset finance and the impact on Indigenous peoples and communities," the interviewee related, adding that "it bubbled up, and there were a lot of people that were mobilizing outside.... Broadly speaking, the mechanics of organizing outside marches to reflect inside policy demands is extremely difficult ... alternative spaces always present a unique challenge because of the merging of all of the diverse interests."

Ultimately, the more radical approach, and that most aligned with vocal Indigenous leadership—the "No REDD" campaign's blanket opposition to REDD and new REDD+ expansions—emerged as the clearest position among justice advocates outside negotiations as well as many inside. As the NGO worker narrated it, "the focus in Copenhagen was largely on the safeguards and the need to respect Indigenous rights. I think in the year following that the increasing trajectory [was] towards something that was very harmful, and the demand was 'we don't want to just tweak on the margins, we just want this to stop; this is not an approach that will work.'"

If the "No REDD" campaign achieved the clarity necessary for mobilization and hewed to the Indigenous and anti-capitalist underpinnings core to the transnational movement, it also suggested that the resonance of principles connected with climate justice can depend on context and positionality. Debate during COP17 over the rights of nature as a frame for coalitional action illustrated additional dimensions of the latter concerns.

Context

On the heels of the massive presence of peoples' organizations in Cochabamba, the apparent appeal of its major statements within the transnational climate justice community in Cancún, and related institutional advocacy by Bolivian representatives in the UN, rights of nature advocates prepared an ambitious calendar for the Durban COP.[76]

---

[76] See, for example, Global Alliance for the Rights of Nature 2015.

Some South African participants at open forums on the UKZN campus, however, considered the rights of nature ill matched to the political and legal context of national and local struggles. Several expressed concern over the potential for conflict between nature's rights and those of poor people. In discussion following a panel session, a scholar-activist issued a sympathetic call for caution:

> I'm worried about the discourse of the rights of nature, and I'm uneasy saying this because it's also great to see so many people linking red and green issues which is something that the environmental justice movement in South Africa has struggled with. But I'm not sure that framing those issues in terms of the rights of nature is going to strengthen their struggles.
>
> For example, we had very, very intense struggles over water and the imposition of technological devices for pre-paid water meters. And in that struggle we have emphasized … firstly the fact that that access to adequate water is a right in terms of our constitution.

Another South African participant expanded on the need to develop an outside climate politics more directly aligned with existing mass movements:

> We've reached the limit of the NGO-ization of the ecological question in multilateral process. In our own contest in South Africa, we really need to think about a mass-based politics. That's why we're sitting in this space. Now I think that how we articulate and how we work alternatives is crucial for that mass-based politics. Right now in South Africa for example, none of the black groups, none of the fronts of struggle, articulates the language of rights of nature, for example. Nobody uses that language – in South African context. We've got to think about, how do we work with the alternatives in our everyday struggle…. We've got to find an intersection point, a common platform, a programmatic platform for mass-based politics.

Others linked this disjuncture with Apartheid histories of forced urbanization, the removal of Blacks from ancestral lands, the absence of well-developed Indigenous identities, and the context of massive contemporary inequality. Foreign activists attending the meetings pointed to related and wider challenges associated with framing alternatives to existing governance in terms of the rights of nature. Who, for instance, would speak for "nature," and how would recognition and protection of its rights respond to conditions of globally uneven human development and ongoing struggles for basic needs in poor, extraction-based national economies?

Later in the meetings, another organizer returned to such concerns, considering how a principle like the rights of nature might be legitimated and mobilized transnationally, and the liabilities of attempting to do so prematurely:

I am very proud that in Ecuador the rights of nature is recognized in our constitution…. but is it a good strategy to build an international movement, a so-called global coalition for the rights of Mother Earth? …

[T]o jump different levels could destroy even a good idea … because rights of nature is probably one of the most revolutionary things in terms of the conceptualization of rights, philosophically, in the past 500 years … if we don't first of all understand locally, in the communities, in our organizations, what we are talking about when we say 'rights of nature,' and … start first of all to compare: 'these Indigenous people in Bolivia … understand rights of nature like this;' and let's compare what is happening with the peoples in Ecuador; let's compare what is happening in Soweto, when you are talking about rights of Mother Earth – and first of all, in a very local area to understand what we are talking about, understand each other among us.

And then – *then* [emphasizes] – we can say, 'okay' …. Only at that moment we can say… 'let's take the rights of nature to confront capitalism, let's take the rights of nature to confront green economy, let's confront rights of nature with the payment for ecosystem services.'

Only at that moment we can jump to a second level and at that moment – maybe, *maybe* – we can start to talk about an international coalition of rights of nature because we have been, for years maybe, knowing what we are talking about, and the idea will be protected by the local level, by the communities.

Like those surrounding REDD, debates on the "outside" over the rights of nature illustrate difficulties in transnational mobilization that extend from the diversity among potential climate justice constituencies and their socio-spatially and historically specific political concerns. In both instances, organizers sought to develop solidarities that could supersede the "militant particularisms" that often motivate such groups within national contexts, in order to facilitate broader alignment and mobilization.[77] The relevance of context takes on political significance of a different order, though, when it lies at the core of differing theories of change.

As organizers debriefed civil society involvement following the Durban COP, they exposed stark differences of orientation toward local contexts and global audiences within the wide range of groups who had planned and participated in the Day of Action march and other events during the negotiations. Some saw those differences, among others, as having limited the impact of civil society as a whole. Sociologist and commentator Ashwin Desai criticized tactics deployed by international NGOs less attuned to local realities, in their complex and socially constitutive particularity[78]:

If people parachute in, do their little stunt, and leave, or get deported [as did several Greenpeace activists attempting a banner hang on a nearby hotel roof] for example,

---

[77] Harvey and Williams 1995.
[78] Desai quoted in Bond 2012, 63–64. See also Hargreaves 2012.

then what do they understand about Durban? What do they understand about the real difficulties of organizing around climate justice? There are real tensions and challenges that people face here, as a stitch between a kind of crony capitalism and African nationalism, but also a kind of rank modernization theory; a "why the fuck shouldn't we have these things?"; "who tells us we shouldn't have cars and TV sets?"

A hundred people were taken to the hospital after the explosion of the Engen refinery [in South Durban], but large swathes of that community are employed by the refineries, so they can't make the move to ask for their closure. And then the climate justice movement asks for them to be closed. What does it mean that people have arrived here, marched and never been to the South Basin? There are 150 smoke stacks. Cancer is everywhere. Nearly every kid carries an asthma pump.

By parachuting in and substituting yourself for local struggles, you won't have a sense of any of that. The way the international NGOs conduct themselves is to adopt the same tactics and strategies everywhere. They have flattened the world and in the process our histories and traditions and our subjectivities.

Such differences of approach, which can also accompany dramatic disparities in resources and access between international NGOs and community groups, understandably arouse the rancor of committed local organizers. They also highlight fundamentally differing orientations to the political, with correspondingly disparate commitments of practice and strategy.

Photographic documentation of major NGOs' protests during COPs, for instance, which are often dramatically staged and visually striking, are staples of international reporting on the meetings. The easy legibility and esthetic appeal that allow the records of those displays of opposition to travel so widely, however, may also enable their subsumption in a familiar narrative, which tells of plodding, ineffective bureaucracy called to account by a few bold and savvy crusaders (often from the Global North). Effective critique of failing governance might come instead from shining a light on the seamier effects of lived, socioecological marginalization reproduced daily just outside the venues of international meetings. Massive, unsightly, and deeply domestic mobilization could well offer a more pointed challenge to assumptions of legitimacy in international process, for participants inside as well as more broadly through media coverage. But as outside and linking organizers are also aware, affected groups are not objects to be manipulated in spectacle. The work of fostering widespread popular mobilization is significantly more challenging, then, in that it demands coherent articulation across differing contexts as well as the material requisites of truly mass participation.

Material Need and the Politics of Presence

Assembling constituencies on the outside is itself a major feat, achieved against significant obstacles. As a first cut, organizers face the material and logistical challenges associated with access to and control of space within which to hold gatherings, and the provision of necessities for the constituents taking part. These concerns are in one sense prior to and yet deeply connected with the constructive work and visible presence that is the focus of the gatherings. They are political as well as practical, and therefore constitute the focus of considerable energy, planning, and resources on the part of organizers. This is because, in general, space for "Peoples' forums" is hard to obtain, distant from negotiations, and subject to state control.

NGO and social movement representatives began planning for a Peoples' Space during the COP in Durban many months before the meetings, with the formation of the C17 committee. Early plans to host gatherings near the COP conference center itself fell through in the summer of 2011, and a scramble ensued to obtain a space large enough to provide the focal point for the outside activities of the many national, regional, and transnational groups planning a presence in the city during the UN talks.[79] As it turned out, more than adequate facilities, including meeting rooms, a café, common spaces and lodgings, were made available at the Howard Campus of UKZN. The distance and terrain, however, separating that location from the convention center, the route of the major march during the COP, and the Speakers Corner necessitated the arrangement and continual re-negotiation of transportation, which remained at a premium throughout the negotiations. Movement between outside and inside spaces, and the visibility of non-accredited constituencies, was thereby made more difficult.

In Copenhagen, Cancún, and Rio the geography of venues was also expansive, though roughly inverted, with the primary "outside" forums located in city centers, and the UN meetings in outskirts or beyond. Such spatial separations reinforce political disconnection at a local scale for the duration of the official meetings. A year after the highly publicized confrontations in Copenhagen, where the conference venue was accessibly by public transit and by foot, the Cancún negotiations were notably more enclaved. Access to the Moon Palace complex from the city, where many less affluent state and observer delegations lodged and the primary outside venue convened, was only possible by buses contracted for that purpose, a thirty-five-minute ride along a stretch of highway closed to all other traffic. UN accreditation badges were verified at the few boarding points in the city, and comprehensive airport-style security screening followed at the conference complex.

---

[79] Bond 2012.

Official UN side events, similarly secured, convened separately in Cancún: in hangers accessible only by a twenty-minute chartered bus ride from the negotiation site itself. The "Cúpula dos Povos" convened during the 2012 Conference on Sustainable Development in an expansive beachfront park in downtown Rio de Janeiro. The park served as the focal point for participation by non-accredited members of Indigenous communities and social movements, an outside node for linking efforts, and the setting for the collaborative drafting of a people's declaration critical of the intergovernmental meetings, as well as the public face for the Conference itself, and promotional stage for loosely related state, private, and public-private initiatives.[80] Large sections of the park were taken up with architecturally appealing temporary constructions devoted, for instance, to clean energy technologies and government programs.

More generally, the use of public space by peoples' groups is subject to the permission and control of local and national states. The comprehensiveness of this control is readily demonstrated by authorities, as it has been around officially sanctioned public demonstrations. Over 900 participants were arrested during the Day of Action march in Copenhagen, for instance.[81] On Durban's Day of Action a fleet of armored police vehicles delayed marchers for some forty-five minutes at the origin point of the agreed route, without explanation, to the consternation of wary participants. A similar contingent of vehicles encircled the municipally approved Speakers Corner in the park across from the conference center later in the day, as members of the DLF congregated there to wait for transportation back to their camp near the university.

Official requirements for approval and performances of control suggest why an unannounced march by the RWA and a few DLF members, while thoroughly peaceful, succeeded in expressing the groups' rejection of the terms of debate inside negotiations much more clearly than did the presentation of their strongly worded memorandum to the UNFCCC Secretary at the conclusion of the sanctioned Day of Action march the following day (see below). That brief illicit action and the enthusiasm with which it was met by other participants in outside gatherings also highlighted the availability and appeal of extra-legal spatial tactics for stakeholders "on the outside." Official management of space and media coverage threatened to enfold the RWA's presence within established institutions and discourses, effacing its more confrontational elements.[82] Examples of those

---

[80] See the official website for the Cúpula dos Povos, and the final declaration of the peoples' summit process that convened there: Government of Brazil 2012; People's Summit for Environmental and Social Justice 2012.

[81] McKie and Zee 2009; Johnson 2012.

[82] See Hargreaves 2012.

tendencies included the Secretary's warm, widely photographed welcome of them on the Day of Action, as well as frequent evocations within the negotiations of extreme vulnerability and promise as represented by African women—often by members of other demographic groups. Illegal mass actions access a far more legible agonistic politics, through the visibility of what Judith Butler terms "bodies on the street" out of place.[83] The presence of the Rural Women's bodies so near the perimeter of the negotiations when they were not expected or condoned constituted a strike against the managed "politics" within which their participation outside the COP was generally contained.

At the same time, the spaces of action and shelter occupied by the RWA, DLF, and other constituency groups point to a tension in these outside climate politics with which justice advocates must contend. On the one hand, it is the presence of such groups where they do not officially belong that foregrounds the exclusionary governmental separation of UN meetings and the increasing ecological precarity these constituencies face, which has mounted unchecked by such meetings for nearly three decades. On the other hand, mitigating that politically salient precarity by ensuring adequate and safe food, lodging, and transportation is in part the responsibility of organizers during the time that members of affected communities take from their own economic, community, and family activities to be present outside negotiations. Balancing these concerns weighs upon organizers, and demands occasional compromise.

The material and logistical challenges associated with obtaining and controlling space, and providing for the needs of assembled constituencies, are therefore closely connected to the political aims of inside-outside linking around securitized forums of global governance. Those aims themselves, and corresponding strategic questions about critical engagement, have also been the subject of serious debate.

Critical Distance and Engagement

Friday evening, Durban: As the last press conferences finish, civil society panelists and reporters stream from briefing rooms to pause in the hall, where throngs of observers have been waiting, eyes glued to screens. Not for the first time, the outcome of the COP remains unresolved at the close of its final scheduled day—the last to which most observers will have access—although the lines of debate between parties are clearly drawn.

---

[83] Butler 2011.

Several observers active in the outside gatherings emerge from one of the press rooms and I follow, exiting the convention center onto a large empty patio enclosed by barricades. It seems unclear to the delegates or to security personnel whether these doors are meant to be secured, but the group clearly means to leave, and no one objects.

Outside the sense of relief is palpable. It is dark, warm, and raining lightly. Sudden contact with actual weather—the immediate experience of what has been the subject of endless calculation and debate these last two weeks—is startling and refreshing.

Finding a route between fences and guards, a small contingent moves across the street to the Speaker's Corner, where a few participants from an Occupy assembly earlier still linger.

Conversations unfold simultaneously: one between two instigators of a large protest earlier today inside the convention center, after which many observers surrendered their badges, a few, apparently, for good; another among a few vocal participants in outside forums including those at the University and the Speakers Corner.

The protestors have passed the remainder of their day here at the Corner. One considers the future. Having refused, non-violently, when asked by security to leave the conference center, he has been told he will not be accredited to a UN meeting again. These have been an important setting of his advocacy work, and it is unclear what he will do next. The other, who did leave when asked, has been barred only from the remainder of this COP. He appears elated, having done today what he came a long way to accomplish. Both, actually, have helped to send a clear and many-voiced message of discontent to their own national delegation, other parties they see as complicit with inaction, the Secretariat, and a small international audience.

In the other group, a seasoned local activist decries what he sees as the cooption of civil society at the COP, a charge he extends even to activities at the University. "We should have had our own meeting," he insists, with bitterness.

The activists who congregated at the Speakers Corner on the final scheduled evening of COP17 shared deep commitment to socio-ecological justice, and deep doubts about the potential of governance-as-usual to deliver it. Their choices differed, however: from critiquing parties during officially sanctioned press conferences within the negotiation venue; to disrupting negotiations in protest and then agreeing to desist, at least for the moment; to refusing to leave a protest and thereby to legitimate the continuing cycle of negotiations that have delivered little; to rejecting engagement with negotiations *in toto*. These suggest the breadth of options among which such critics must choose.

For all the labor and resources expended on stakeholder presence and representation at COPs and other summits like Rio+20, such strategies and tactics have also been the subject of disagreement amongst thinkers and activists dedicated to socio-ecological justice. To what degree, runs a prominent line of questioning, are these forums and actions truly "outside" of post-political mega-events like the COPs? To what degree, by contrast, can the engagement of stakeholders with those events and the participation of social justice organizations in them be understood as merely legitimating the very governmental arrangements and outcomes activists would decry?[84]

The challenge of engagement without cooption was neatly encapsulated in the culminating moment of the Day of Action march in Durban. A column of several thousand marchers had assembled on the morning of December 3 at a park north of downtown, and traveled slowly southward through the center city, passing the International Convention Center (the ICC, in which the COP convened) before ending in a party on the beachfront.[85] Between the Speakers Corner and the ICC, the marchers paused. There, representatives of several participating groups, gathered at the front of the column were greeted by UNFCCC Secretary Christina Figueres and South African Environment Minister Maite Nkoana-Mashabane amidst a swarm of microphones and cameras. The marchers presented a memorandum apparently "from civil society" to the two diplomats, who thanked them for their work and promised to bring their voices into the COP.[86] To critics, Figueres and Nkoana-Mashabane were thereby able to incorporate the potentially insurgent energies of the marching groups, with the apparent backing of their full mass, into a media-ready narrative of their unity and the institutional embrace of stakeholder participation. That representation obscured the real interests of many of the marchers in alternative processes and outcomes, widely felt opposition to South African as well as international climate change responses (and other governmental

---

[84] Cf. Wainwright and Mann 2018.

[85] News sources placed the number of marchers between 8000 and 20,000; for example, The Telegraph 2011; Vidal 2011.

[86] See official host country documentation: COP17 CMP7 2011.

initiatives), and the differing specific demands of the groups present, some of which would remain unspoken on the "inside."[87]

Such concerns about the silencing of frontline voices are tied to others that, for many observers both inside and out, cast the legitimacy of global climate governance in doubt. First are its distinctly unimpressive outcomes to date, in terms of environmental protection or broadly shared "sustainable development." To this is added the steady drumbeat of emerging science, which repeatedly heightens activists' understanding of the urgency with which alternative energy futures must be embraced in order to forestall the worst—and most unequal—social consequences of global warming. Many read these factors as indicating that strategies of costly, frustrating, and time-consuming engagement are misplaced, arguing instead for directly oppositional approaches akin to those adopted in earlier anti-globalization protests.[88] On the other hand, insiders and linking groups highlight the potential for increasing ecological damage and exacerbated threats to marginalized communities that could accrue through the machinations of governance in the absence of critical pressure of civil society, however circumscribed its presence or diluted its contributions might be.

These issues are examined with special urgency in reflexive debates among organizers, in part because theories and designs of governance often attribute a prominent role to civil society involvement in helping to ensure legitimacy, and because critically oriented leaders are often painfully aware of the potential for their involvement to defang insurgent movements through representation, despite their intentions. Such a discussion on the relative merits of engaging with or withdrawing from the COP amidst dire scientific predictions and scant progress within negotiations occurred in a panel discussion midway through the meetings at UKZN. An audience member broached the topic:

> Mainstream sources are saying that this is the last chance to stabilize climate change under two degrees warming, and that may sound abstract but.... If that's true... And if they don't come to an agreement here to radically cut greenhouse gases, over a hundred million people are going to die on this continent alone.
>
> So, [those] who say that we should be inside, you may very well be right, that over time we could take this space back.... But I'm curious if that's a relevant argument given the time we have. I'm concerned that the energy that goes into that could also go into just shutting this thing down.

---

[87] See, for example, Hargreaves analysis of the exchange. Hargreaves 2012.

[88] See discussion of this general strategy of comparatively greater engagement on the part of radical climate justice activists in Mueller 2012. Differing forms of engagement in governance and recent direct action-based mobilization strategies are taken up in Chap. 6.

Organizers on the panel responded, sharing the questioner's concerns of urgency and the need for robust action "outside," but rejecting the impulse toward the latter as a singular focus, and also, implicitly, the notion of "shutting down" the COP. "You got to do everything that's necessary," offered one, "we got our foot in both areas: we're involved in non-violent direct action; we're taking it to the street." "I think you absolutely have it right," responded another, "it's a question of time," adding "you have to use it absolutely, by any means necessary, and that means that you have to operate at every single level, and in every single space." Going on, the second organizer depicted the UNFCCC as a target of consequence within a necessarily broader strategy of simultaneous and varied engagement:

> [T]hey're really negotiating on who has the right to live in this future. Who has the right to food. Who has the right to stay where they stay, and who is going to be forced to be a climate refugee.... [P]eople say "Oh, this UNFCCC, oh we shouldn't be there. That's not our space, we're the movements, and we should be in other spaces." Well, I'm sure they'd love for us to be in other spaces, they'd love for us not to be there....
> [T]here is a power imbalance.... For me, fundamentally, the UNFCCC is about that. It's not a dead space, it's a contested space. It's a space where we may not, at the moment, have political power or leverage, but it's not space that we can ignore.... This is not the WTO, this is not something that if you crash and you said "please don't make a global agreement on trade," and we looked for some alternative [sic]. That's not the reality.

The speaker went on to describe functions many critical organizations play inside the negotiations: offering alternative narratives, and seeking to hold parties accountable to their marginalized constituents amidst international pressure tactics:

> If we're not there, don't think other organizations won't be there.... [O]rganizations that propose the very things that we fight....
> There are hundreds if not thousands of journalists that congregate around those talks. And when you actually speak to most of them ... they have very little idea of what's going on in terms of the negotiations.... [T]hey're sold a framework, a framing, by governments. If we're not there providing an alternative view, an alternative analysis of what's going on in terms of the negotiations, then that space closes and those voices aren't heard....
> [T]he African [countries] group is the most critical group in these negotiations.... [A]nd we know, exactly as we knew in Copenhagen and Cancun, that [African] environment ministers, and the ministers – the prime ministers – will get phone calls from capitols in Europe, from London and Paris and New York and Washington, and they'll say, 'weaken your position.' Unless we counter that, unless we say, 'you as an elite are not beholden to the elites in the North, you're beholden to our people, our priorities' ... they're going to make those closed-door agreements.

For critical civil society groups, like that of this organizer, who came to dominate the debate at the panel session, the original mandate to UNFCCC parties, to safeguard communities and ecologies, remains crucial even while the practice of governance entails real threats of increasing harm and injustice connected with climate change and coordinated responses to it. Those threats therefore demand continuing critical engagement along the tactical lines he described, to the extent such groups' transnational structures and capacities enable it. In this vision, part of the responsibility of representative civil society at global governance fora is to mind the store: tracking governmental leaders' fidelity, or lack thereof, to the demands of their constituents, who are unable to present themselves in international debates. "That's not to say that the UNFCCC is the place where we will win," he concluded, "it's not, it's not ... until we have built our movements so powerful on the outside, so powerful at a local and national level, that they are holding our governments accountable before they come to the COP."

Conclusion

Climate justice has emerged and gained traction as a movement globally in large part through the work of activists, analysts, and affected groups at the margins of international meetings and in peoples' spaces outside them. That work is fundamentally connective, in that it consists of assembling affected and allied groups in inclusive dialogue; articulating climate change with multiple social movement traditions, lived experiences, and forms of knowledge; and presenting the resulting grievances and transformative visions so that they may impact on governmental debates. It has produced unprecedented political alliances and shifted the discourse of climate change in civil society and, though incompletely, within governance circles.

The gestation of these developments in alternative summits and demonstrations during major intergovernmental meetings is the result of formidable effort in the face of material and political obstacles, and the reflexive negotiation of both internal and external tensions. In particular, interactions "on the outside" suggest that a broad-based transnational politics of climate justice requires the grounding of positions in the conceptions and practices of many stakeholder groups, as well as the ability and inclination of members to present themselves. As linking advocates' comments suggested during meetings outside COP17, those requirements in turn imply prior and simultaneous work at local and national levels.

Arguably, a growing number of initiatives and organizations exemplify such geographically and socially grounded efforts, connecting the global socio-ecological processes of climate change with the core concerns and resources of existing political communities or inspiring the emergence of new ones. The promise in these efforts lies in nurturing wider political ecological consciousness, solidarity, and collective action amidst distinct ongoing histories and struggles. The following chapter examines two of these, both set in the US.

Bibliography

ActionAid. (2009, December). *Rich Countries' "Climate Debt" and How They Can Repay It.* Retrieved from https://www.actionaid.org.uk/sites/default/files/doc_lib/updated_climate_debt_briefing_december_200.pdf

ActionAid. (2011). *Fiddling with Soil Carbon Markets While Africa Burns....* Retrieved March 17, 2015, from http://www.actionaidusa.org/shared/fiddling-soil-carbon-markets-while-africa-burns-0

ActionAid, & 16 others. (2010a, November). *Climate Justice Briefs #1: Climate Debt.* Retrieved from http://www.ips-dc.org/wp-content/uploads/2010/12/1-Climate-debt.pdf

ActionAid, & 16 others. (2010b, November). *Climate Justice Briefs #12: Human Rights and Climate Justice.* Retrieved from http://www.whatnext.org/resources/Publications/Climate-justice-briefs_full-setA4.pdf

Banktrack. (2011). BankTrack.org—news—Bankrolling Climate Change. Retrieved March 5, 2013, from http://www.banktrack.org/show/news/bankrolling_climate_change

Bond, P. (2011). *The Politics of Climate Justice.* London: Verso and Pietermaritzburg: University of KwaZulu-Natal Press.

Bond, P. (2012). Durban's Conference of Polluters, Market Failure and Critic Failure. *Ephemera, 12,* 42.

Bond, P., & Dorsey, M. K. (2010). Anatomies of Environmental Knowledge & Resistance: Diverse Climate Justice Movements and Waning Eco-Neoliberalism. *Journal of Australian Political Economy, 66,* 286–316.

Bote, T. (2011, November 30). *Rural Women Farmers Speak Out on Climate Change.* Retrieved March 17, 2015, from http://www.actionaidusa.org/zimbabwe/2011/11/rural-women-farmers-speak-out-climate-change

Building Bridges Collective. (2010). *Space for Movement: Reflections from Bolivia on Climate Justice, Social Movements and the State.* Retrieved from http://www.climatecollective.org/media/uploads/resources/space_for_movement1.pdf

Büscher, B., & Arsel, M. (2012). Nature™ Inc.: Changes and Continuities in Neoliberal Conservation and Market-Based Environmental Policy. *Development and Change, 43*(1), 53–78.

Butler, J. (2011, September). *Bodies in Alliance and the Politics of the Street.* Presented at the The State of Things, organized by the Office for Contemporary Art Norway, Venice. Retrieved from http://suebellyank.com/wp-content/uploads/2011/11/ola-reader-full.pdf

Canadian Youth Delegation, joined by 137 delegates and 21 observer organizations. (2012). *Fossil Foul: Letter to Christiana Figueres.* Retrieved March 4, 2013, from http://fossil-foul.tumblr.com/

Centre for Civil Society. (n.d.-a). *Home—Events & Action.* Retrieved March 17, 2015, from http://ccs.ukzn.ac.za/default.asp?2,68

Centre for Civil Society. (n.d.-b). *Wolpe Lectures and Reviews 2011.* Retrieved March 16, 2015, from http://ccs.ukzn.ac.za/default.asp?11,22,5,2668#Rights%20of%20Nature

Chan, K. M., et al. (2016). Opinion: Why Protect Nature? Rethinking Values and the Environment. *Proceedings of the National Academy of Sciences, 113*(6), 1462–1465.

Chatterton, P., Featherstone, D., & Routledge, P. (2013). Articulating Climate Justice in Copenhagen: Antagonism, the Commons, and Solidarity. *Antipode, 45*(3), 602–620. https://doi.org/10.1111/j.1467-8330.2012.01025.x.

Climate Justice Now! (2007). *Founding Statement.* Center for Civil Society. Available at: http://ccs.ukzn.ac.za/default.asp?4,80,5,2381

Climate Justice Now! (2008). *Climate Justice Now! Statement.* Carbon Trade Watch. Available at: http://www.carbontradewatch.org/index.php?option=com_content&task=view&id=227&Itemid=95

Climate Justice Now! (2010). *CJN! Network Members (as at November 2010).* Available at: https://web.archive.org/web/20150901040407/http://www.climate-justice-now.org:80/category/climate-justice-movement/cjn-members/

Climate Justice Now! (2011). *2011 COP17 Succumbs to Climate Apartheid!* Retrieved from http://www.climate-justice-now.org/2011-cop17-succumbs-to-climate-apartheid-anti-dote-is-cochabamba-peoples%E2%80%99-agreement/

COP17 CMP7. (2011). *Multimedia—Images.* Retrieved March 13, 2015, from http://www.cop17-cmp7durban.com/en/multimedia/image-gallery.html

Derman, B. B. (2014). Climate Governance, Justice, and Transnational Civil Society. *Climate Policy, 14*(1), 23–41.

Dismantle Corporate Power. (2012, October 17). *United Nations: Who Wants to Go Through the Revolving Door?* Retrieved March 4, 2013, from http://www.stopcorporateimpunity.org/?p=2213

EJOLT (Environmental Organizations, Liabilities, and Trade). (2013). *Ecological Debt.* Retrieved from http://www.ejolt.org/2013/05/ecological-debt/

Environmental Justice Climate Change Initiative. (2002). *10 Principles for Just Climate Change Policies in the United States [EJCC].* Retrieved from http://www.ejcc.org/about/

Environmental Justice Leadership Forum on Climate Change. (2015). *Environmental Justice Leadership Forum on Climate Change.* Retrieved March 17, 2015, from http://www.ejleadershipforum.org/

FOEI (Friends of the Earth International). (2012). *Reclaim the UN from Corporate Capture.* Retrieved from http://www.foei.org/resources/publications/publications-by-subject/economic-justice-resisting-neoliberalism-publications/reclaim-the-un-from-corporate-capture/

Friman, M. (2007). *Historical Responsibility in the UNFCCC.* Retrieved from http://www.diva-portal.org/smash/record.jsf?pid=diva2:233089

Gersmann, H., & Vidal, J. (2011, November 28). *Q&A: Durban COP17 Climate Talks.* Retrieved March 9, 2013, from http://www.guardian.co.uk/environment/2011/nov/28/durban-cop17-climate-talks

Global Alliance for the Rights of Nature. (2015). *Rights of Nature Advocates at COP17 Durban, South Africa.* Available at: http://therightsofnature.org/rights-of-nature-cop17-durban/

Global Justice Ecology Project. (2015). *About Climate Justice.* Retrieved March 17, 2015, from http://globaljusticeecology.org/climate-justice/

Global Witness. (2011). *Making the Forest Sector Transparent: Annual Transparency Report 2009.* Global Witness. Retrieved March 5, 2013, from http://www.globalwitness.org/library/making-forest-sector-transparent-annual-transparency-report-2009

Goodman, A. (2011, December 14). On Climate Change, the Message Is Simple: Get It Done. *The Guardian.* Available at: http://www.theguardian.com/commentisfree/cifamerica/2011/dec/14/durban-climate-change-conference-2011

Government of Bolivia. (2009). *Commitments for Annex I Parties Under Paragraph 1(b) (i) of the Bali Action Plan: Evaluating Developed Countries Historical Climate Debt to Developing Countries.* Retrieved from http://unfccc.int/resource/docs/2009/awglca6/eng/misc04p01.pdf#page=44

Government of Brazil. (2012). *Cúpula dos Povos—Rio + 20.* Retrieved March 13, 2015, from http://www.rio20.gov.br/clientes/rio20/rio20/sobre_a_rio_mais_20/o-que-e-cupula-dos-povos.html

Grassroots Global Justice Alliance. (2011, December 4). *Rural Women's Assembly of Southern Africa Statement to COP17 Leaders.* Retrieved March 17, 2015, from http://ggjalliance.org/node/897

Gudynas, E. (2011). Buen Vivir: Today's Tomorrow. *Development, 54*(4), 441–447. https://doi.org/10.1057/dev.2011.86.

Gudynas, E. (2017). Value, Growth, Development: South American Lessons for a New Ecopolitics. *Capitalism Nature Socialism.* https://doi.org/10.1080/10455752.2017.1372502.

Hargreaves, S. (2012, May). *COP 17 and Civil Society: The Centre Did Not Hold* (Institute for Global Dialogue, Occassional Paper No. 64). Retrieved from http://www.igd.org.za/publications/occasional-papers

Hart, G. (2013). Gramsci, Geography, and the Languages of Populism. In E. Michael, H. Gillian, K. Stefan, & L. Alex (Eds.), *Gramsci: Space, Nature, Politics* (pp. 301–320). West Sussex: Wiley-Blackwell.

Harvey, D., & Williams, R. (1995). Militant Particularism and Global Ambition: The Conceptual Politics of Place, Space, and Environment in the Work of Raymond Williams. *Social Text,* (42), 69–98.

Heinrich Böll Foundation. (2011). *Mobilising Climate Finance for Africa.* Retrieved March 10, 2013, from http://www.et.boell.org/web/113-266.html

International Council on Human Rights Policy. (2009). *Human Rights and Climate Change.* Cambridge University Press. Retrieved from http://eprints.lse.ac.uk/id/eprint/25596

Johnson, C. (2012, January 26). *Denmark: Court Upholds Decision in Mass Arrests Case* [web page]. Retrieved March 18, 2015, from http://www.loc.gov/lawweb/servlet/lloc_news?disp3_l205402960_text

Jubilee South. (2010). *Towards a Peoples' Agenda on Climate Finance*. Retrieved from http://apmdd.org/component/phocadownload/category/2-ecological-debt-environmental-justice-and-climate-change?download=8:towards-a-peoples-agenda-on-climate-finance-a-part-of-our-platform-for-climate-justice

Keck, M. E., & Sikkink, K. (1998). *Activists Beyond Borders: Advocacy Networks in International Politics*. Ithaca: Cornell University Press.

Keck, M. E., & Sikkink, K. (1999). Transnational Advocacy Networks in International and Regional Politics. *International Social Science Journal, 51*(159), 89–101.

Kodili, B. C. (2011). *7000 km in 17 days to COP17*. Retrieved March 16, 2015, from http://www.actionaid.se/en/activista/shared/7000km-17-days-cop17

Kovel, J. (2007). *The Enemy of Nature: The End of Capitalism or the End of the World?* London: Zed Books.

La Via Campesina. (2010a, December 3). *The International Caravan of La Via Campesina Advances for Cancun*. Retrieved March 16, 2015, from http://www.viacampesina.org/en/index.php/actions-and-events-mainmenu-26/-climate-change-and-agrofuels-main-menu-75/958-the-international-caravan-of-la-via-campesina-advances-for-cancun

La Via Campesina. (2010b, December 10). *Via Campesina Declaration in Cancún: The People Hold Thousands of Solutions in Their Hands*. Retrieved March 16, 2015, from http://www.viacampesina.org/en/index.php/actions-and-events-mainmenu-26/-climate-change-and-agrofuels-mainmenu-75/984-via-campesina-declaration-in-cancun-the-people-hold-thousands-of-solutions-in-their-hands

La Via Campesina. (2011a). *La Vía Campesina Declaration in Cancun*. Retrieved March 16, 2015, from http://viacampesina.org/en/index.php/actions-and-events-mainmenu-26/-climate-change-and-agrofuels-mainmenu-75/992-la-via-campesina-declaration-in-cancun

La Via Campesina. (2011b). *Via Campesina at COP17 in Durban: Industrial Agriculture Heats Up the Planet. Farmers Are Cooling It Down!* Retrieved March 16, 2015, from http://viacampesina.org/en/index.php/actions-and-events-mainmenu-26/-climate-change-and-agrofuels-mainmenu-75/1126-via-campesina-at-cop17-in-durban-indus-trial-agriculture-heats-up-the-planet-farmers-are-cooling-it-down

Lohmann, L. (2008). Carbon Trading, Climate Justice and the Production of Ignorance: Ten Examples. *Development, 51*(3), 359–365. https://doi.org/10.1057/dev.2008.27.

Mary Robinson Foundation—Climate Justice. (2011, November 18). *Climate Campaign Travelling Across Africa to COP17*. Retrieved March 16, 2015, from http://www.mrfcj.org/news/climate-campaign-travelling-across-africa-to-cop17.html

Massey, D. B. (2005). *For Space*. London: SAGE.

McKie, R., & Zee, B. van der. (2009, December 12). *Hundreds Arrested in Copenhagen as Green Protest March Leads to Violence*. Retrieved March 18, 2015, from http://www.theguardian.com/environment/2009/dec/13/hundreds-arrested-in-copenhagen-violence

Mueller, T. (2012). The People's Climate Summit in Cochabamba: A Tragedy in Three Acts. *Ephemera, 12*, 70.

Occupy Cop17. (2014). *Occupy COP17*. Retrieved March 16, 2015, from https://www.facebook.com/occupyCOP17

One Million Climate Jobs. (2011). *Climate Jobs Booklet 2011*. Retrieved March 16, 2015, from http://climatejobs.org.za/

Onyango, O. (2011, November 2). *Addis Ababa Meeting Consolidates African Unity on Climate Change*. Retrieved November 18, 2019, from Trans African Caravan of Hope website: http://transafricancaravanofhope.blogspot.com/2011/11/addis-ababa-meeting-consolidates.html

Paterson, M. (2011). Selling Carbon: From International Climate Regime to Global Carbon Market. In J. S. Dryzek, R. B. Norgaard, & D. Schlosberg (Eds.), *The Oxford Handbook of Climate Change and Society* (pp. 611–624). Oxford: Oxford University Press.

People's Summit for Environmental and Social Justice. (2012, July 20). *Final Declaration of the People's Summit at Rio+20*. Retrieved from http://www.internationalrivers.org/files/attached-files/rio20_peoplessummit_eng.pdf

REDD-Monitor. (2019). *REDD-Monitor*. Retrieved from http://www.redd-monitor.org/

Roberts, J. T., & Parks, B. C. (2009). Ecologically Unequal Exchange, Ecological Debt, and Climate Justice: The History and Implications of Three Related Ideas for a New Social Movement. *International Journal of Comparative Sociology, 50*(3–4), 385–409.

Robin Hood Tax. (2010). *A Global Financial Transaction Tax for Climate Funding: Investing in Our Collective Future*. Retrieved from http://www.robinhoodtax.org/sites/default/files/RHTC%2520Climate%2520Paper_0.pdf

Routledge, P. (2017). *Space Invaders: Radical Geographies of Protest*. London: Pluto Press.

Routledge, P., & Cumbers, A. (2009). *Global Justice Networks: Geographies of Transnational Solidarity*. Manchester: Manchester University Press.

Rural Women's Assembly. (2011). *Memorandum from the Rural Women's Assembly to President Zuma and Minister Nkoane-Mashabane*. Retrieved from https://ruralwomensassembly.wordpress.com/cop17/memorandum/

Spash, C. (2011). Carbon Trading: A Critique. In J. Dryzek, R. Norgaard, & D. Schlosberg (Eds.), *The Oxford Handbook of Climate Change and Society* (pp. 550–560). Oxford: Oxford University Press.

Stamp Out Poverty. (2009). *Assessing the Alternatives*. Retrieved from http://tilz.tearfund.org/~/media/files/tilz/research/sopassessing_web.pdf

Stamp Out Poverty & Institute for Development Studies. (2011. December). *Climate Finance*. Retrieved from http://www.gci.org.uk/Documents/Climate_Finance_.pdf

Stand With Africa. (2012). *Stand With Africa: Act Now for Climate Justice*. Retrieved March 17, 2015, from https://www.facebook.com/pages/Stand-With-Africa/258982300799452

Swyngedouw, E. (2005). Governance Innovation and the Citizen: The Janus Face of Governance-Beyond-the-State. *Urban Studies, 42*(11), 1991–2006.

Swyngedouw, E. (2007). Impossible "Sustainability" and the Postpolitical Condition. In R. Krueger & D. Gibbs (Eds.), *The Sustainable Development Paradox: Urban Political Economy in the United States and Europe* (pp. 13–40). New York: Guilford Press.

Swyngedouw, E. (2010). Apocalypse Forever? Post-Political Populism and the Spectre of Climate Change. *Theory, Culture & Society, 27*(2–3), 213–232.

Takver. (2011, December 10). *OccupyCOP: Hundreds Protest Inside UN Climate Venue in Durban as Talks Draw to a Close*. Retrieved March 17, 2015, from http://www.indybay.org/newsitems/2011/12/10/18702334.php

The Telegraph. (2011, December 9). *Durban Climate Change Conference 2011 Latest*. Retrieved from http://www.telegraph.co.uk/news/earth/environment/climat-echange/8916405/Durban-Climate-Change-Conference-2011-latest.html

Third World Network. (2009). *Climate Debt: A Primer*. Retrieved from http://www.twn.my/announcement/sign-on.letter_climate.dept.htm

UNCSD (United Nations Conference on Sustainable Development). (2011). *Green Economy*. Retrieved March 16, 2015, from http://www.uncsd2012.org/greeneconomy.html

UNFCCC (United Nations Framework Convention on Climate Change). (2019). *REDD+ Home*. Retrieved June 26, 2019, from https://redd.unfccc.int/

United Nations. (2011). *International Year for People of African Descent 2011*. Retrieved March 17, 2015, from http://www.un.org/en/events/iypad2011/global.shtml

Vidal, J. (2011, December 5). *Durban Climate Talks: Day Eight Diary*. Retrieved March 13, 2015, from http://www.theguardian.com/environment/blog/2011/dec/05/durban-climate-talks-day-eight-diary

Vidal, J. (2012, December 4). *Doha Climate Conference Diary: Youth Activists Bring Energy and Urgency*. Retrieved March 5, 2013, from http://www.guardian.co.uk/environment/blog/2012/dec/04/doha-climate-conference-diary

Wainwright, J., & Mann, G. (2018). *Climate Leviathan: A Political Theory of Our Planetary Future*. London/Brooklyn: Verso.

World Development Movement & Jubilee Debt Campaign. (2009, November). *The Climate Debt Crisis: Why Paying Our Dues Is Essential for Tackling Climate Change*. Retrieved from http://slettgjelda.no/assets/docs/Rapporter/climatedebtcrisis1.pdf

WPCCC. (2010a). *People's Agreement*. Retrieved March 4, 2013, from https://pwccc.wordpress.com/support/

WPCCC. (2010b). *Rights of Mother Earth*. Retrieved March 4, 2013, from http://pwccc.wordpress.com/programa/

Grounding Climate Justice

5

This chapter examines two efforts to link the Anthropocene ties and uneven geographies of climate change with the social and environmental justice concerns of pre-existing and emerging political communities and garner influence for those concerns through established national and subnational channels. The Environmental and Climate Justice Program (ECJP) of the National Association for the Advancement of Colored People (NAACP) highlights the connections by which climate change disproportionately affects people of color, in the US and abroad. Doing so allows the ECJP to leverage the established constituency and advocacy capacities of the larger organization. The Cowboy Indian Alliance (or New CIA), a coalition of Native American and white activists, formed out of independent efforts to prevent a pipeline project across the Midwestern US. The Alliance has pursued a culturally and geographically resonant politics of protection, highlighting the ability of shared concern to underpin solidarity in the face of a long legacy of conflict, and helping to repeatedly block the pipeline.

The ECJP and New CIA's work suggests empirical models of mobilization and advocacy that respond to imperatives sensed by a variety of participants and analysts in climate justice and other global movements against social and ecological marginalization.[1] Analyzing the operation of "global justice networks," for example, Routledge and Cumbers argue that "though global events and networks are important, arguably more time and resources should be spent on networking locally

---

[1] For example, Featherstone 2008; Routledge and Cumbers 2009; Routledge 2011; Bond 2012; Mueller 2012.

© The Author(s) 2020

B. B. Derman, *Struggles for Climate Justice*,
https://doi.org/10.1007/978-3-030-27965-3_5

and nationally," and Routledge calls further for "constructing more effective *grounded* resistance to injustice and responses to climate change" in particular.[2] Indeed, the challenges of mobilizing around COPs and the recalcitrance of global climate governance to the influence of people's movements described in the preceding chapters have inspired some insiders to assert the priority of local and national initiatives in preparing the ground for, legitimating, or even obviating transnational ones.[3]

Transnational efforts have succeeded in exposing the global ties of responsibility that link emitters and affected people, and providing the analytical and social frameworks for similarly extensive solidarities among dispersed and diverse groups (see Chap. 4). Work like that of the ECJP and New CIA offers a crucial complement to their efforts by giving equal or greater emphasis to historically, socially, and geographically specific conditions. These "grounded" efforts differ, though, from exclusionary movements for identity- and place-based social and environmental protection in that wider patterns and solidaristic ties are constitutive in the ECJP and New CIA's advocacy and activisms.[4] The full spatial and temporal extents of global warming's industrial ecology are logically necessary for the ECJP's analysis of climate change as an issue of civil rights as understood in the US context, for instance, just as confronting centuries of violence and oppression plays a key role in cementing and legitimating the New CIA's coalitional opposition to contemporary fossil energy infrastructure and its locally concentrated hazards. Together, the two efforts discussed here suggest important additional dimensions of a thoroughgoing socio-ecological politics based in the recognition, construction, and mobilization of connections.

The ECJP: Political Community and Advocacy Infrastructure

> When folks think about climate change, what often comes to mind are melting ice caps and suffering polar bears. Historically, American society has failed to make the connection in terms of the direct impact of environmental injustices, including climate change, on our own lives, families, and communities.... Climate Change and other environmental injustices are about US.[5]

---

[2] Routledge and Cumbers 2009, 215; Routledge 2011, 394, emphasis in original.

[3] See, for example, Bond 2012; Mueller 2012; Ecosocialist Horizons 2011.

[4] See, for example, Harvey 1996.

[5] NAACP 2018a.

Founded in 1909, the NAACP is the most long-lived organizational actor associated with the movement for racial justice in the US.[6] Best known for its role in realizing the judicial and legislative civil rights milestones of the 1950s and 1960s, the group's voter registration and mobilization efforts remain influential.[7] Recent reporting places its membership at approximately 500,000, across more than 2000 chapters.[8] Since 2009, "climate justice" has emerged and risen in prominence among its areas of advocacy. The ECJP links the social dimensions of climate change with the core concerns of the NAACP's comparatively well-defined and established constituency. Those situating analyses in turn support the ECJP in leveraging existing channels of advocacy: mobilizing the organizational and political infrastructure through which the larger organization has long represented Americans of color in legal and policy debate.

The ECJP's analysis links the NAACP's core thematic commitments, which are not commonly defined in environmental or ecological terms, with the differential impacts experienced by people of color at moments of production, impacts, and regulation of climate change. In the ECJP's research reports and campaigns, places, bodies, and biographies join spatial and demographic patterns as the elements through which the socio-ecological ties and socio-spatial disparities of anthropogenic climate change are demonstrated and framed as issues of civil rights, health, and economic injustice.

By tying the continued social marginality of Black Americans with the industrial and ecological processes of the fossil fuel economy, the ECJP socializes, historicizes, and politicizes those processes in the context of ongoing social struggle. Its connective analyses thereby provide the basis for climate justice advocacy in the American civil rights tradition, and for solidarity-building beyond the historical geographic purview of that movement, with people of color facing differential climate impacts worldwide and other allies. In the domestic context specifically, framing climate change-related impacts on Black Americans in terms of disproportionate health, education, and economic effects renders those impacts as issues for local, state, and national policy intervention, and thereby as topics for advocacy and comment utilizing the NAACP's established public profile and expertise. Drawing on those resources, the ECJP can challenge the political disenfranchisement that contributes to the continuation of racialized local concentrations of socio-ecological injustice: highlighting, for

---

[6] Berg 2007; Sullivan 2009.
[7] Berg 2007; Goluboff 2007.
[8] Shipp 2018.

instance, energy companies' decisions about plant location, fuel, and technology, which disproportionately affect communities of color.[9]

Making Climate Connections

The ECJP's connective analyses demonstrate the socially consequential ties linking fossil fuel production and resulting climate change impacts to the lives and life chances of Black Americans—for NAACP members as well as decision makers and broader audiences. In an interview, an ECJP leader described beginning the process with trainings held with the larger organization's regional assemblies:

> We have regional meetings, and do training on our civil rights agenda. So, when I did the "Climate Justice 101" training … [participants] said "Oh, I thought this was going to be about climate and workplace discrimination." And … "I thought this was going to be about a climate of justice in the world." It made me really see how far people were coming even with the term. But on the other end of that statement was, "wow, we really get how this connects to everything," because it was very story-based.

One example of the ECJP's story-based technique is the Women of Color for Climate Justice Road Tour, which resulted in a series of short online videos[10] profiling individual women who "are experiencing differential impact, are involved in local self-reliance campaigns and are undertaking efforts to resist negative environmental developments."[11] Many videos document individual women's experiences of living with and fighting coal-fired power plants in or near their neighborhoods and workplaces. Some speak from the experience of weathering Hurricanes Katrina and Rita and later BP's Gulf oil spill. Showing the women themselves, often in their homes or outside the plants they oppose, the videos foreground the embodied, place-based realities underlying wider patterns of disproportionate industrial location and disaster risk in poor neighborhoods and communities of color, with particular attention to the gendering of health impacts and resistance efforts.

Video sets juxtaposing the local consequences of fossil fuel energy production with the "downstream" impacts of rising tides and intensifying storms in coastal communities exemplify a more general analytical strategy of the ECJP. This consists of highlighting the demographically patterned, differential social impacts associated with multiple moments in the corporate-driven, industrial-ecological

---

[9] See, for example, NAACP et al. 2012.

[10] NAACP n.d.

[11] Patterson 2009a.

production of anthropogenic climate change. These moments include the mining, refining, combustion, and waste disposal steps in the fossil energy production process, in addition to its climatic consequences in sea level rise and severe weather.[12] Examining the processes of human development, agricultural industrialization, and incarceration as well thickens and grounds the ECJP's analyses of climate-related injustice. The organizer continued:

> [Using narrative] was really how I was able to illustrate the connections between everything from the drivers of climate change, to coal power plants, to these other greenhouse-gas emitting facilities, to talking about how this affects us on the other side with Hurricane Katrina. … [T]he whole picture around not only climate change itself, but the corporate entities that are driving climate change and are also driving those markets that affect us in all these different ways.
> And even talking about criminal justice … one of [the NAACP's] foundational issues … is that we have a school-to-prison pipeline. … [W]hen half our kids have asthma and half have ADHD, both of which are affected by toxins in the air, how 70% of us live in communities and counties in violation of air pollution standards, and then we wonder why so many of us are incarcerated. Because the rates show that if you're not reading at grade level by grade 3, you're more likely to be involved in the criminal justice system.
> So these are things that people see every day, but they don't necessarily make those ties.

More conventional quantitative and spatial analytic work complements the ECJP's narrative approach. Its research report *Coal Blooded: Putting Profits Before People*, for instance, analyzed the dominance of coal-fired energy in the US and demonstrated its disproportionate impacts on poor communities and people of color at local, national, and global scales through air pollution and climate change.[13] *Coal Blooded* ranks the "Environmental Justice Performance" of 378 US coal plants, combining individual plant scores with demographic analysis to demonstrate just how unequally the health impacts of coal energy production are socialized, reproducing the marginalization of Americans of color.[14] The report also assigns a

---

[12] See, for example, NAACP 2013; NAACP et al. 2012. For instance, their 2013 Just Energy Policies report (discussed later) states: "Not only do low-income neighborhoods and communities of color suffer more of the direct health, educational, and economic consequences of these facilities, but also devastating natural disasters such as Hurricanes Katrina and Sandy, along with rising food prices and water shortages, harm low-income people and people of color disproportionately partly due to pre-existing vulnerabilities (NAACP 2013, 5)." In the report *Coal Blooded: Putting Profits Before People*, the group discusses these as the intersection of environmental and climate justice. NAACP et al. (2012).

[13] NAACP et al. 2012.

[14] The report calculates that: "[a]pproximately two million Americans live within three miles of one of [the] 12 [worst] plants and the average per capita income of these nearby residents

ranking to the fifty-nine energy companies operating coal plants in the US, and calls for the closure of seventy-five "failing plants."[15]

Like the transnational analyses of socio-ecological connection discussed in the Chap. 4, the ECJP's work is sensitive to the economic context of fossil fuel energy production, and the potential consequences of mitigation and renewable alternatives on poor communities. Thus, like the UK and South African Climate Jobs campaigns, NAACP critiques of coal have been tightly tied to proposals for increasing economic justice and opportunity based in an emergent clean energy economy.[16] This emphasis also links the ECJP's work with the current programmatic priorities of the larger organization, which include economic justice.[17] Its analytical consideration of economy-ecology connections has led the ECJP to expand its focus on energy policy, as in the 2013 report *Just Energy Policies: Reducing Pollution and Creating Jobs*, which argues, "[p]owering our nation, protecting the environment and empowering communities of color are not mutually exclusive concepts. In fact they are essential and interrelated calls to action."[18] Thus, as the report states, in addition to addressing the environmental injustices of a fossil-fueled economy chronicled in *Coal Blooded* and elsewhere, "energy policy can create real public benefits, including millions of good green-collar jobs and building an inclusive green economy strong enough to lift people out of poverty."[19]

Common Cause and Local Context

Analyzing disproportionate climate-related impacts on poor communities and people of color, and working within the geographically extensive network of the NAACP's US organizational structure as well as its national and transnational partnerships, plays an important part in constructing the ECJP's constituency-based

---

is \$14,626 (compared with the U.S. average of \$21,587). Approximately 76% of these nearby residents are people of color (NAACP et al. 2012, 29)." A 2002 report determined that 68% of African-Americans lived within the toxic release zone (designated by a thirty-mile buffer) of a coal-fired power plant. See Black Leadership Forum et al. (2002).

[15] NAACP et al. 2012.

[16] For example, NAACP 2010a; NAACP et al. 2012. The *Coal Blooded* report argues that closing the seventy-five most environmentally unjust coal-fired energy plants would cut energy supply by a mere 8%, which could be recuperated through increased energy conservation and renewable energy production. NAACP et al. (2012, 58).

[17] NAACP 2010a, b.

[18] NAACP 2013, 535.

[19] Ibid, 5.

and solidaristic politics of climate justice. At the same time, the initiative's interventions (e.g. in addressing the worst-ranked coal-fired power plants identified in *Coal Blooded*) are also attuned to the role of local social relations in individual communities' place-based struggles.

The group's internally oriented educational materials and externally focused policy advocacy both link storm impacts in the coastal US with the toxic shadow of coal production in the Midwest.[20] Its online splash page captures this analysis of shared experience[21]:

> Environmental injustice is about people in Detroit, Ohio, Chicago, Memphis, Kansas City, and elsewhere who have died and others who are chronically ill due to exposure to toxins from coal fired power plants and other toxic facilities.
>
> Climate change is about the increase in the severity of storms which means that storms like Sandy and Isaac, which devastated communities from Boston to Biloxi, will become more of the norm. Our sisters and brothers in the Bahamas, as well as Inuit communities in Kivalina, Alaska, and communities in Thibodaux, Louisiana and beyond, who will be losing their homes to rising sea levels in the coming few years.
>
> Climate change and environmental injustice are about sisters and brothers from West Virginia to Tennessee who are breathing toxic ash from blasting for mountain top removal.

Collecting local and state-level analyses in the videos and reports, and sharing them both online and at trainings and teach-ins around the country, constructs solidarity across such sites and constituencies, as well as providing opportunities for sharing models and stories of success. *Just Energy Policy*, for instance, includes a table presenting promising policy approaches by state, while *Coal Blooded* features summaries of successful plant closure and retro-fitting campaigns; both reports and the ECJP *Climate Justice Toolkit* include model letters and statements that local and state activists can adopt as templates.[22]

Complementing such efforts at broad solidarity and action, the Program's site-specific response work exemplifies its attention to differences in the dynamics of power and opportunity among local places. The interviewee quoted earlier described the group's strategic attention to context in terms of communities' economic necessities:

> We have all these tools in our tool box … for example in [community A, the power plant] has no revenue for the community, they don't bring any jobs to the community…. So [community leaders] have developed two strategies. One is that they

---

[20] See, for example, Patterson 2011; NAACP et al. 2012, 30.
[21] NAACP 2018a.
[22] NAACP 2010a, 2013; NAACP et al. 2012.

developed a local ordinance that they're trying to get passed through city council....
On the other hand, they're really pushing hard with a big PR campaign making sure
that people know how bad this is, how the company isn't doing anything: basically
calling out the company.

In a place like [community B], if you're not directly employed [at the plant] then
your business depends on the fact that everyone [else] is employed there. So that's a
place where the best we can hope for is pollution control.... So what we end up doing
and how we end up doing it is very much dependent on where.

The ECJP also partners with other national and community-based environmental
and climate justice groups, and leverages the institutional relationships of the larger
NAACP organization. The comprehensive analysis presented in *Coal Blooded*, for
instance, was authored collaboratively with the Indigenous Environmental
Network, Little Village Environmental Justice Organization (LVEJO), and
Rainforest Action Network.[23] Similarly, the Women of Color for Climate Justice
Road Tour and the video library it launched began in partnership with Women of
Color United, Women's Environment and Development Organization, and the
Environmental Justice and Climate Change Initiative.[24] The ECJP's website also
includes a page linking to national, state-level, youth, and campus-based "resource
organizations."[25] The NAACP's extensive institutional connections further amplify
its climate justice work, and link it with additional intellectual resources and political actors. For instance, in 2010 ECJP staff convened a forum of representatives
from Historically Black Colleges and Universities (HBCUs) and US governmental
agencies "to discuss the engagement of HBCUs in planning and executing a research agenda on the [BP] oil drilling disaster, as well as ongoing sustainability in
the Gulf region."[26]

The Program has also reached further to develop and strengthen transnational
alliances, reporting on and organizing both "inside" and "outside" activities during
UN climate change negotiations.[27] The ECJP leadership coordinated several events
with the Pan-African Climate Justice Alliance before and during COP17, for instance.[28] These efforts expand the NAACP's global orientation even as they ground

---

[23] NAACP et al. 2012, 1. Several individuals involved in climate justice analysis and mobilization are also thanked for their contributions.

[24] Patterson 2009a.

[25] NAACP 2018b.

[26] Patterson 2010b.

[27] For example, Patterson 2009b.

[28] Patterson and Njamnshi 2011. In a jointly organized panel discussion in Durban, members
of the two groups discussed struggles and means of resistance, including advocacy and legal

the transnational discourse of climate justice, through solidaristic analyses of common drivers and issues affecting people of color in the US and abroad.

This Human and Civil Rights Issue

The ECJP's connective ecological and social analyses underpin its position that climate change poses civil as well as human rights issues: the production and impact of climate change affect different communities in different ways, but examined in aggregate each of its moments differentially harm people of color across the US and the globe.[29] In an early summary statement for *The Root*, director Jacqueline Patterson linked the diverse modalities of climate change impact with existing vulnerabilities in Black communities, and formulated political responses in terms of the democratically driven realization of rights[30]:

> Whether it is sea-level rise causing dislocation; severe storms taking homes, lives and communities; Black children and families starving or sick from respiratory illnesses or exposure to carcinogenic toxins; children missing school or performing poorly due

---

levers, in the US and in various African country settings. The Initiative also added to its online video collection during the COP, posting testimonials from African women facing climate impacts. NAACP (n.d.).

[29] NAACP 2010a, 2018a. The ECJP's *Climate Justice Toolkit*, an early comprehensive statement prepared for NAACP constituents' use toward these ends, enumerates implications for rights (NAACP 2010a, 4):
HUMAN RIGHTS VIOLATIONS
Climate change negatively impacts the following human rights:

- Right to Self Determination
- Right to Safe and Healthy Work Conditions
- Right to Highest Standard of Physical and Mental Health
- Right to Food
- Right to a Decent Living Condition
- Equal Rights Between Men and Women
- Right of Youth and Children to be Free From Exploitation

CIVIL RIGHTS VIOLATIONS

- Climate change negatively impacts the following [civil] rights
- Ensuring peoples' physical integrity and safety
- Protection from discrimination on grounds such as gender, religion, race, sexual orientation, national origin, age, immigrant status, etc.
- Equal access to health care, education, culture, etc.

[30] Patterson 2010a.

to resulting illness; or heat exposure resulting in illness or death; African-American communities are often starting from a place of substandard school systems, compromised access to quality health care, as well as job, housing or other vulnerability which makes facing these challenges even more impactful than they would be on a person or community with more resources and access to quality services....

From those we have elected to office as decision makers and duty bearers, we must demand real reductions in emissions; representation in policy and program design, planning, implementation and evaluation processes; and reparations for what has been taken from us through the excesses of the many, through provision of resources; and preservation and upholding of our civil rights as constituents and our human rights as people.

Thus, according to the group's statement of purpose, "The NAACP Environmental and Climate Justice Program was created to support community leadership in addressing this human and civil rights issue."[31] Framing climate change as an issue of human rights links the ECJP's work with a major thread of transnational advocacy for climate justice. Understanding it as an issue of domestically recognized civil rights in turn supports the mobilization of the NAACP's considerable organizational capacities for giving voice to the concerns of communities of color in advocacy for legal protection and policy change.

Organizational Resources

Policy is a major focus for the ECJP, which both complements and benefits from its other central mission of education. The NAACP's established organizational infrastructure helps the ECJP address the challenges of political disenfranchisement, which tend to isolate groups most affected by climate change and critical civil society actors from representative politics and consequential decision-making processes. The group's policy work, which is facilitated by the connective analyses and practices described earlier, mobilizes both the broad constituency and the institutional capacities of the larger organization. Accordingly, its repertoire for advocacy efforts includes detailed policy analysis and recommendations, mobilizing voters and branch organizations, occasional direct access to agency officials, legal mobilization, and public comment across a variety of platforms.

The Program's *Just Energy Policy* report, and to a somewhat lesser extent *Coal Blooded* and the *Climate Justice Toolkit*, present detailed rights- and economic justice-based analyses of existing government policy and corporate practice

---

[31] NAACP 2018a.

regarding energy, health, environment, and minority opportunities. Each document offers comprehensive recommendations for policy change, grounded in existing provisions within model state-level provisions where those exist.[32] Further, the Program's connective analyses shape more targeted and time-sensitive policy statements and campaigns, articulated in press releases, action alerts, and organization-wide resolutions.[33]

Action alerts, press releases, resolutions, and toolkits also facilitate independent actions by individual NAACP members and chapters. Toolkits and action alerts typically include templates for letters that constituents can send to their elected representatives and ideas for local campaigns and educational programs. Voter education and mobilization also takes place through action alerts, blog posts, and the NAACP's yearly national- and state-level "legislative report cards," which track elected officials' performance on matters of concern to the organization.[34] With these tools, then, the ECJP can work to mobilize the broader NAACP constituency and empower units across its organizational structure. It also strives to ground its advocacy in constituents' experiences and amplify their voices through advocacy. The organizer quoted previously described one of the forms this linking takes, which also illustrates the close tie between the ECJP's education and policy work as these involve NAACP members and units:

> [We've] been doing teach-ins around the country [on the EPA's Mercury Air Toxics Standards rulemaking process]. At the end of the teach-in, we all sit and do testimony ... and then the state conference president might write a letter to the editor or the policy maker, and will include the quotes from their constituents that say "this is why we need strong policy."

Through processes like this one, the organization's membership provides resources for advocacy, while shaping it in ways that ensure and reinforce legitimacy.

In addition to mobilizing the NAACP's large member base, the ECJP leverages its well-established political and legal advocacy arms. These sometimes allow ECJP leaders direct contact with agency officials around climate change impacts affecting Americans of color. Examples include an inter-agency briefing and a gathering of agencies and educational institutions focused, respectively, on response planning and research needs following BP's Gulf oil spill.[35] Legal mobilization, a historically important NAACP strategy, also forms an aspect of the ECJP's

---

[32] For example, NAACP 2013.
[33] For example, NAACP 2011a, b, 2018c.
[34] For example, Patterson 2010c; NAACP 2018c, d.
[35] Shannon 2010; Patterson 2010b.

work, as in a 2014 appeals court battle for which it partnered with other environmental justice and environmental organizations to successfully defend the Environmental Protection Agency (EPA) Mercury Air Toxics Standards.[36]

Finally, ECJP and NAACP leaders contribute public comment on climate justice-relevant policy issues as authors and sources across a variety of progressive, environmental, Black, and mainstream media outlets.[37] In each of these ways, the ECJP leverages the NAACP's organizational capacities to give voice to analyses of climate injustices and pursue more appropriate responses through institutional channels.

The ECJP's socio-ecologically connective analyses provide the content for its outreach efforts around the country, and for internal and more widely distributed publications through which an established political community is hailed as the subject of a climate justice politics based in civil and human rights. The policy work of the initiative leverages this wider constituency, the organization's advocacy infrastructure, and the Program's socio-ecological rights-based analysis to influence processes in government at a variety of scales. The ECJP's linking of climate change with the established politics, policy, and law of racial justice, which has supported them in accessing social mobilization and advocacy infrastructure connected with those concerns, exemplifies the use of analytical articulation, or "translation" in Hart's terms, to expand opportunities for broad-based as well as institutional forms of Anthropocene-era socio-ecological advocacy.[38]

Many of the ECJP's analyses and practices resonate with those developed and deployed by other organizations, including groups focused on energy sovereignty, Indigenous people's rights, and economic justice.[39] Its emphasis on the local legibility of climate injustice and suitability of responses is widely shared by US groups coalescing out of or joining in movements for climate justice.[40] The leader quoted previously highlighted the importance and insight of community-based groups, in particular, and the Program's mission has evolved to explicitly prioritize support for them.[41] The ECJP's focus on educating and mobilizing a widening

---

[36] Earthjustice 2014.

[37] See, for example, Jealous 2010, 2013; Anft 2013; Fontaine 2014; Mock and Patterson 2014; Tincher 2014.

[38] Cf. Hart 2013; Massey and Rustin 2013.

[39] For example, Our Power Campaign 2016; Indigenous Environmental Network 2018a; Poor People's Campaign 2018.

[40] See, for example, Stephenson 2015; Our Power Campaign 2016.

[41] NAACP 2018a.

range of constituencies for climate justice is also shared by civil society groups active across a range of sites and scales. Together with such like-minded community, national, and transnational organizations, the ECJP is beginning to fill important lacunae in US awareness and political action around the social inequalities of climate change.

The articulation of emerging needs and transformative demands with pre-existing discursive and institutional frames is, however, often slow, difficult, and subject to limits, and like any organization, the ECJP faces opportunities and constraints particular to its own history as well as the socio-political context in which it operates. Arguably, the popular understanding and the legal and policy framing of civil rights already established by the NAACP and other groups present unusual opportunities, at least in the US context, for articulation with the social inequalities that climate change and fossil energy use tend to re-inscribe. Economic justice, in comparison, has lacked comparable institutionalization and, until recently, comparable salience in the US.[42] Class-based mobilization for climate justice may therefore present even greater challenges than efforts centered on civil rights and racial justice in that context.

The ECJP's own pivot from analysis to advocacy depends not only on the cultural, legal, and policy infrastructure surrounding civil rights, but on the NAACP's unique membership, institutional ties, and expertise. To some extent, its broad membership and name recognition constitute assets in education and mobilization. On the other hand, the paucity of direct legal avenues for environmental justice in the US[43] dictates that despite the NAACP's more widely recognized capacity for effective litigation, much of the ECJP's work must proceed by other means: adding motivation, perhaps, to its extra-legal analytical, mobilization, and alliance-building efforts.

While the ECJP's work illustrates the role of socio-ecological analysis in facilitating the articulation of climate justice with existing movement energies and resources, the New CIA's efforts suggest ways in which the material bases of fossil fuel energy systems affect particular places and the groups that call them home, and the new solidarities such impacts can inspire.

---

[42] Scheingold 1974; Chomsky 2013; Booker 2018. The neglect of economic justice as a social value in the US may be changing, however, as indicated by the staying power of the idea of the "1%" associated with the Occupy movement and the recent re-emergence of a poor people's movement connected with the late work of Dr. Martin Luther King Jr.

[43] Herbert et al. 2013.

People vs. Pipelines

> [T]he Keystone XL tar sands pipeline has brought communities together like few causes in our history.[44]

At the same time that more and more populations and polities have begun to grapple with the anthropogenic evolution of the earth's climate, the fossil fuel industry has had to confront shifts in the geology of oil availability. Although a ready oversupply necessitated the industry's maintenance of market scarcity through much of the twentieth century, firms have now for decades faced the diminishing production of established wells, coupled with a falling rate of discovery of new oil fields. Industry and observer estimates still place reserves at levels more than adequate to shift the earth's climate system beyond the bounds human and other species depend on, but the value of firms is based on the reserves to which they claim access, and demand for energy within the current fossil fuel-based system continues to rise.[45] In combination, these factors have driven the industry into an era of "unconventional" oil exploration and production, in which sources like "tight" or "shale" oil and tar sands have become crucial sources, to which firms devote significant investment both for extraction and transportation.

Climate change is not the only reason for concern, as these unconventional sources gain market share and efforts to produce them expand. In more immediate environmental terms, tar sands oil is a particularly costly and dangerous form of fuel, beginning at the moment of extraction. The oil in tar sands, called bitumen, is unusually viscous: comparable in consistency to molasses or even Silly Putty. Bitumen occurs densely mixed with sediment in large patches beneath top soil and sometimes in deeper deposits. Tar sands extraction differs from "conventional" oil operations, where drilling accesses liquid crude oil trapped under pressure in deep cavities, to be controlled by caps and valves at wellheads and then pumped to the surface as pressure within the deposit decreases. Near-surface bitumen extraction is, instead, akin to strip mining, eventually requiring the destruction of soil and vegetation overlaying the entire deposit. Deeper troves can be steam-heated and pumped to the surface. While the latter method obviously requires significant energy to bring the fuel-laden material to the surface, in both situations separating the bitumen from the accompanying sediments also demands large quantities of energy and water, and produces polluted water and sediment tailings. It is for this

---

[44] Reject and Protect 2014a.
[45] Mitchell 2011.

reason that only amidst decreasing reserves of conventional oil and spikes in price, tar sands deposits became cost-effective sources of supply.[46]

Transporting bitumen, often through pipelines, entails risks as well. Leaks are not uncommon in pipelines generally, but spill risks from bitumen lines seem to exceed those associated with conventional crude. Because bitumen is denser, it is heated, pressurized, and mixed with hydrocarbon diluents to facilitate flow within the line (becoming diluted bitumen, or dilbit).[47] The National Resources Defense Council (NRDC) documented a higher rate of spills in pipelines carrying dilbit than those transporting conventional oil.[48] Dilbit leaks may be harder to detect because of the pipeline technologies involved, and more costly and disruptive to remediate. When the Enbridge corporation's Line 6B pipeline leaked into Michigan's Kalamazoo river in 2010, for instance, its Edmonton-based operators initially assumed that an air bubble in the pipe had caused the lowered pressure reported by its monitoring technology, so they increased pressure to clear the bubble, spilling more oil. A Michigan utility worker identified the spill and notified the company, which was able to close off the pipeline seventeen hours into the spill.[49] By then, approximately one million gallons of dilbit had entered the river, becoming one of the largest inland spills in US history. As the diluents in the dilbit evaporated, the heavier bitumen sank, requiring extensive, ecologically damaging dredging operations, over some forty miles of the river, to remove. Primary cleanup activities continued into 2014. A National Transportation Safety Board report called the spill the costliest onshore recovery effort ever in the US. Enbridge settled damages with the EPA and Department of Justice in 2014 for 177 million dollars, estimating its cleanup costs at 1.2 billion.[50] By June of 2018, public agencies and environmental groups had verified a small fraction of the cleanup; verification and testing for impacts continues.[51]

As the Kalamazoo River spill demonstrated, water bodies near pipelines are highly vulnerable to leaks, and their own extents and flows enable leaked material to travel widely. The TC Energy, formerly TransCanada Corporation's, Keystone XL (or KXL) pipeline project, originally proposed in 2008, would create a more direct link between the extensive Athabaskan tar sands region of Alberta, Canada, and terminals located in

[46] Mitchell 2011; Minnesota Public Radio 2018.
[47] Minnesota Public Radio 2018.
[48] National Resources Defense Council 2013.
[49] National Transportation Safety Board 2012.
[50] Ellison 2014.
[51] Michigan Department of Environmental Quality 2018.

Illinois and Texas.[52] It has been delayed repeatedly, in part because its initial proposed route included the environmentally sensitive Sand Hills region of Nebraska. A revised route would skirt some vulnerable areas, but grave concerns remain around the potential effects of spills and leakage into the Ogallala Aquifer, a massive unground water reserve that supports agricultural production in eight "breadbasket" US states, over which the pipeline would also travel.[53]

Oil and gas pipelines have provided the focal point for extremely energetic public protests in recent years. Activists target pipelines for a variety of reasons, building impressive and often influential networks of organized grassroots and institutional opposition. First Nations groups from the areas surrounding the Athabaskan tar sands were the first to mount significant opposition to bitumen mining there. They are joined by Indigenous activists on both sides of the US-Canada border and further away, in opposing tar sands and other fossil fuel extraction and transportation projects including KXL, the Dakota Access Pipeline (DAPL), and Enbridge's Line 4 replacement in Minnesota[54] because of disproportionate impacts in their own communities as well as wider effects.

The destruction of forests and soils atop tar sands reserves, massive water consumption, and release of polluted tailings from processing all directly affect Indigenous lands. These impacts have been associated with diseased fish and game and rampant cancer in First Nations communities near production sites.[55] For some, these localized environmental effects are closely linked to a more general disharmony among people and nature encompassing anthropogenic climate change, which also disproportionately impacts Indigenous peoples in North America and around the globe.[56] In the US particularly, these "environmental" impacts represent some of the socio-ecological dimensions of oppression through colonization, genocide, removal, and re-education, and the repeated abrogation of treaty, human, and Indigenous rights guaranteeing social protection and community control or use of ancestral lands, water, and sacred sites, both within and beyond areas over which tribes maintain sovereignty.[57]

---

[52] TransCanada Corporation 2018.

[53] Brown 2014.

[54] Owe Aku 2014; Adler 2015; Minnesota Public Radio 2018.

[55] Indigenous Environmental Network n.d.; Great Plains Tar Sands Resistance n.d.

[56] For example, Protect the Sacred n.d.; Wildcat 2013.

[57] See Protect the Sacred n.d. Indigenous peoples' rights in land in the US are the subject of complex, contested, and often forgotten legalities. Legal rights to the use of wild species on

Though not every effort has succeeded in preventing fossil fuel projects, their impacts have grown, and they have had other effects in the development or renewal of alliances and identities. As the breadth of the concerns already listed suggests, organizing against KXL and other projects has lent impetus to new and revivified ties of solidarity among tribes and First Nations, which extend across the US-Canada border and well beyond the paths of proposed or existing pipelines. In some cases, groups once pitted against each other have come together to oppose those projects.[58]

At the same time, as author Anton Treuer notes, protests like those ignited by the Standing Rock Sioux Tribe's opposition to DAPL showed that organization against pipelines and other projects have also played a role in redefining identities for many Native people.[59] For some, roles connected with these protests, like "water protector" and "defender of the sacred," have become important elements, adding specific, historically resonant but contemporary meaning. Former vice-presidential candidate Winona LaDuke captured the notions of re- and newly emerging identities as well as the development of widening solidarities, stating that, at Standing Rock, "we remembered who we were," and "people we do not know came to stand with us."[60]

Non-Native activists were among those who came, as "No DAPL" protests grew both at Standing Rock and in solidarity elsewhere. Thousands of protestors ultimately joined Sioux leaders in the 2016 encampment against DAPL. Years earlier, grassroots opposition along the path of KXL had begun to build on the legacy of previous alliances between neighboring Native and rural white communities. In the 1980s, for instance, similar coalitions organized against uranium and munitions testing in North Dakota's Black Hills.[61] Often effective in blocking such projects, these alliances also sometimes succeeded in improving troubled relations between their constituent communities.[62] In 2011, tribal leaders in crucial areas along the proposed path of KXL began organizing with white ranchers and landowners who would also be affected by pipeline development. These groups

---

ancestral and ceded territory outside reservations, for instance, are often abrogated and widely misunderstood. See, for example, Silvern 1999; Minnesota Public Radio 2018.

[58] Moe 2014a; Minnesota Public Radio 2018.

[59] Treuer 2018, quoted in Minnesota Public Radio 2018.

[60] Minnesota Public Radio 2018.

[61] Aronoff 2017; Grossman 2017.

[62] Grossman (2017) examines these, including early efforts against KXL, in his study of "unlikely" Indigenous–rural white alliances for environmental protection.

linked their energies around what they recognized as shared local interests, in ways that would help to provide needed grounding for a national climate movement in the US, which had also identified the pipeline as a key target. They revived the intentionally provocative name of earlier, similar efforts, calling their coalition the Cowboy Indian Alliance, and drew on other aspects of their differing traditions, both to develop solidarity and gain wide attention to their cause. The anti-KXL Cowboy Indian Alliance—sometimes called the New CIA, or CIA III—continued to fight, and block, the pipeline, even as the political context in which its efforts first gained traction changed dramatically.

The Cowboy Indian Alliance

Members and allied activists have recounted the trajectory of the Cowboy Indian Alliance against KXL, in their own documentation and that of journalists and scholar-activist Zoltan Grossman.[63]

Like the NAACP before the ECJP, the Cowboy Indian Alliance did not convene to address the anthropogenic climate change toward which KXL would contribute. Indigenous leaders involved in the coalition have repeatedly emphasized their communities' treaty rights, which would be abrogated by the pipeline, as well as the material threats it would pose to land and water. The white landowners and ranchers with which they partnered were initially motivated to defend their property rights against takings by eminent domain, and prevent the environmental impacts of pipeline spills on their land-based livelihoods. (Parcels owned by some more vocal whites were excluded in early, minor revisions to the pipeline route.)[64] Leader of the anti-KXL group Bold Nebraska, Jane Kleeb, who came to play a central role in the Alliance, has told of coming to the pipeline conflict after organizing in other issue areas, including youth voting and health care.[65] As whites joined Native-led resistance efforts, the two groups recognized that international corporate interests threatened the primary values each held, in different ways, in the integrity of the land over which the pipeline would pass.[66]

Despite these common interests, of course, as Grossman notes, "the recurring irony of the alliance is that the ranchers have been fighting against the corporate theft of their property, that their forebears had themselves stolen from Native

---

[63] See especially Moe 2013, 2014a, c; Reject and Protect 2014d; Grossman 2017.

[64] Grossman 2017.

[65] Adler 2015.

[66] Tejada and Catlin 2013; for example, Protect the Sacred n.d.; Grossman 2017.

people."[67] As in the earlier coalitions, that history, and its contemporary legacy in racism, political disenfranchisement, and simmering local conflicts, posed challenges to the group's solidarity that could not successfully be avoided. Alliance leaders did not try to avoid them, intentionally working toward mutual understanding and unity in the shadow of violence and dispossession, by creating opportunities to develop social connections between members through practices like sharing food and stories, as well as joint participation in some of the spiritual aspects of Indigenous members' relationship with land and water. As journalist Kristen Moe described it, "[n]aturally, the alliance organizes on conference calls and on smart phones, but they make time for in-person gatherings, some of which last for several days, where time is given to sitting around and just talking [.... t]elling stories, introducing their grandkids, spending time out on the land" at spiritual camps and other events.[68]

One of the New CIA's principle conveners, Dakota/Nakota elder Faith Spotted Eagle, has said, "[a]t some point we have to backtrack and unpack our bags, and begin to figure out what happened between us as neighbors," naming the holocaust and dispossession Natives suffered at the hands of settlers. To function as members of the Alliance, white ranchers and landowners had to come to understand and respect Native experiences and treaty rights. Ultimately, in Spotted Eagles' judgment, they did "get it," at least in part because they now faced a form of dispossession as well.[69] Ranchers and farmers spoke passionately, in early meetings, of protecting the land in which their families had invested generations of labor, for children and grandchildren.[70] But the similar experiences of whites and Natives in the area are also layered historically: the farm that now belongs to the Tanderups, for example, another white family involved in the Alliance, sits not only on the KXL's proposed path but on the Ponca Trail of Tears, along which members of the Native group, which now opposes KXL, passed during removal from their homelands by the white federal government.[71] For Alliance members, the 2013 signing of the International Treaty to Protect the Sacred, by multiple Native groups as well as some whites, symbolized new understanding and respect. As Native groups resolved among themselves to put aside old enmities in favor of joint opposition to

---

[67] Grossman 2017, 182.

[68] Moe 2014b.

[69] Moe 2014a, c; Grossman 2017.

[70] Minnesota Public Radio 2018.

[71] Abourezk 2018.

KXL, whites pledged to honor treaty rights that have been repeatedly shunted aside or disparaged by the American settler state and local landowners alike.[72]

As Indigenous Environmental Network (IEN) executive director Tom Goldtooth suggested, alliance is itself a crucial strategy of the New CIA: "it's about political power," he noted.[73] As the group worked to understand each other, find common ground, and pool their different reservoirs of political opportunity and influence, its diversity drew wide attention to its cause, reflecting the experiences of similar previous coalitions with which key Native organizers were familiar.[74] But the coalitional and identity-based aspects of the New CIA's political power are also meaningful beyond their instrumental utility against KXL. 350.org First Nations organizer Clayton Thomas-Muller stated that "we need to develop an organizing framework that effectively addresses racism, oppression, misogyny and colonialism," noting that, in this respect, the New CIA "represents an important step towards reconciling America's bloody colonial history."[75] The Alliance and its allies' definitively regional opposition efforts also provide evidence that rural white Americans can retain independent voices amidst national political debates often dominated by elites who either ignore them or claim to speak on their behalf.[76] For Kleeb, the Alliance's successes illustrate that "we actually do matter, in middle America."[77] It may be that the cultural, spiritual, and geographical touchstones by which the New CIA built its framework of reconciliation and reached wider exposure suggest means by which ecological politics can be grounded more deeply in social context, and thereby become both more diverse and more broadly meaningful.

Articulating Water, Land, and Climate

Climate activists certainly took note of the potentially potent grounding the New CIA and regional groups like Bold Nebraska offered for their cause. Experienced environmental organizer Kenny Bruno noted that pipeline opposition offered a tangible focus, which had previously been largely missing from the US climate

---

[72] Grossman 2017.
[73] Quoted in Democracy Now 2017.
[74] Grossman 2017.
[75] Quoted in Moe 2014b.
[76] Aronoff 2017.
[77] Ablow 2017.

movement.[78] Local and regional anti-pipeline fighters, who stood to be directly affected by fossil fuel infrastructure development, brought even more immediacy to climate-oriented campaigns. Kleeb's group suggested that people from "middle America" *were* concerned, and blocking the pipeline itself would be a clearly identifiable win. Environmental author and 350.org founder Bill McKibben contacted Kleeb to enlist her group's participation in the first large-scale anti-KXL protests planned for Washington, DC, in 2011. In fact, the regional pipeline fighters and climate activists both sought to broaden their audience and impact: national groups shared resources and expertise, but, as Kleeb put it, "they also needed our stories."[79] Recognizing new opportunities in partnering with nationally oriented climate activists, Bold Nebraska farmers and ranchers made the journey, notching their first arrests for civil disobedience, and helping to launch nation-wide opposition to KXL.[80]

Even as its reach grew, however, the forefront of Bold Nebraska's activism—and that of the New CIA—remained the need to protect a place, along with the livelihoods and meanings that inhere in it. Into 2014, as the Obama administration repeatedly postponed a decision on KXL under pressure from environmentalists, New CIA members and allies balanced relationship-building and regional actions.[81] Indigenous coalition Idle No More and Bold Nebraska organized a joint demonstration outside the Nebraska state capital in January 2013, opposing Governor Dave Heineman's approval of the pipeline.[82] In November, Native groups and white neighbors gathered for a spiritual camp on the Tanderup farm.[83] In April 2014, the Alliance created a crop art image representing a Native warrior, a cowboy, the sun, and the water of the aquifer with the inscription "HEARTLAND #NoKXL," covering an eighty-acre cornfield outside Neligh, Nebraska.[84]

Later that month, in the Alliance's largest public action yet, a contingent of Native people and white farmers and ranchers rode into Washington, DC, on horseback, clad in cowboy hats and traditional regalia (Fig. 5.1). They lit a

---

[78] Minnesota Public Radio 2018.

[79] Minnesota Public Radio 2018.

[80] Adler 2015; Minnesota Public Radio 2018.

[81] Because it would cross the US-Canada international border, the KXL route requires federal executive branch approval.

[82] Abourezk 2013.

[83] Moe 2013.

[84] See Associate Press 2014. Collaborating artist John Quigley and Bold Nebraska estimated the image the largest crop art ever created. Hefflinger 2014.

Fig. 5.1 Cowboy Indian Alliance members bring shared demands to Washington, DC during Reject and Protect, in April 2014. (Photo by Chip Somodevilla/Getty Images)

ceremonial fire, erected tipis, and occupied an encampment on the National Mall for a six-day action against the pipeline dubbed "Reject and Protect."[85] Leaders and members of the Ponca and Lakota nations, Idle No More, and Bold Nebraska spoke at the gathering; 350.org assisted with logistics. The Alliance and its allies had linked KXL with climate and community in their call for presence and solidarity at the protest[86]:

> The Cowboy and Indian Alliance and our allies invite folks from across the country to visit the tipi camp on the National Mall and participate in the actions.... Join us in showing the strength of our communities. We call upon President Obama to take this historic step in rejecting Keystone XL in order to protect our land, water and climate.

---

[85] Reject and Protect took place on April 22–27, 2014, in Washington, DC. Reject and Protect 2014d; Capriccioso 2014; Roland 2014. Planned prior to the Administration's latest postponement, the camp was scheduled strategically to influence the earlier timeline for its decision.

[86] Reject and Protect 2014d.

An online ride board coordinated trips from across the country. As many as 5000 people joined in a mass march on the penultimate day of the gathering,[87] as media coverage highlighting the more photogenic displays of Native-settler identity and solidarity reached around the globe.[88]

During Reject and Protect, pipeline protest took artistic and spiritual forms as well as more common spatial ones, including the occupation of the Mall and other public spaces. As one observer wrote: "[h]onoring the sacred took precedent over all other activities at the encampment."[89] Each day began with a water ceremony, in which a draught drawn from the Ogallala was blessed and poured over the soil. A ceremonial fire burned throughout the gathering. Events on the Mall included dance and musical performances, collective paintings, stories, interfaith prayer, and exhibitions of photojournalism and film about KXL and the tar sands. Elsewhere, participants held a prayer meeting outside Secretary of State John Kerry's house, and a round dance blocking an intersection near government and lobbyists' offices. The Alliance and other Native leaders gifted a hand-painted tipi to the Smithsonian's National Museum of the American Indian, honoring President Obama, their coalition, and pipeline opposition. One panel of the tipi was decorated with thumb and handprints of visitors to the camp.[90]

In press releases, on the Mall, and in a meeting with members of the Obama administration, representatives of the Alliance spoke of their coming together around common values, experiences of disenfranchisement, and concerns for the future. After the gathering, participants reflected on an increasing understanding among Natives and whites, and strengthened commitment to their common cause.[91] Kleeb received an email from the White House, saying, "OK, you've got our attention."[92] The administration had postponed a decision on KXL less than a week prior to the CIA protest, saying it would allocate additional time to consider the "unprecedented" 2.5 million public comments it had received.[93] In 2015, the administration decided against approval of the pipeline, based partly on what it described as "extensive public outreach" and consultation.[94]

---

[87] Reject and Protect 2014c; Moe 2014a.
[88] Grossman 2017.
[89] Roland 2014.
[90] Reject and Protect 2014b, c.
[91] Moe 2014b.
[92] Adler 2015.
[93] Mufson 2014.
[94] Chappell 2015.

Almost immediately upon assuming the US presidency in January 2017, Donald Trump signed a memorandum inviting the pipeline company to "re-submit its application" to build KXL. In March he issued the company a permit to complete the project.[95] Opponents recognized, however, that other avenues existed to block the pipeline, and continued their opposition with renewed vigor, responding in many variations on IEN lead organizer Dallas Goldtooth's assessment: "[w]e've stopped the toxic Keystone XL pipeline once and we will do it again."[96]

Like its anti-pipeline allies, the New CIA began making headlines again after the change in administrations, sustaining and mounting new actions evoking their ties to the land. Each year, beginning in 2014, the group and its supporters have planted a strain of corn sacred to the Ponca Tribe on a portion of the Tanderup farm.[97] In June 2018, the family deeded the tract back to the Ponca, creating the first Native land on the pipeline route, with potential new legal hurdles in the way of construction.[98] The New CIA also began the "Solar XL" campaign, with 350.org and other allies, erecting solar panels and a solar-powered barn on the pipeline route. The renewable energy installation, built by community members, would have to be dismantled to make way for the corporate dilbit pipeline project.[99]

The Alliance and its partners have also continued their fight through official processes and electoral politics. In August 2017, the Alliance marched in Lincoln, in anticipation of a Nebraska public utilities commission ruling on the pipeline route. When the commission ruled in favor of construction, opponents took it to court. Kleeb assumed leadership of the Nebraska Democratic Party, and Spotted Eagle, who had received an electoral vote for president in 2016, mounted a run for congress. Pipeline fighters have begun to run in "hyper-local" elections as well, seeking seats on commissions and in district offices: low profile but crucial settings for approval or rejection of infrastructure projects.[100] In November 2018, a federal court ruled that the Trump administration erred in approving KXL without further evaluation of its environmental impact.[101] The latter half of 2019 saw a major ruling in favor of construction, as well as a major spill on TC Energy's existing Keystone line. The possibility of oil price disruptions, and proposals to expand existing Canada-US pipelines beyond the volume necessary to move supplies have also cast doubt on the project's viability.[102]

---

[95] Mufson and Eilperin 2017; Jamieson and Vaughan 2017.

[96] Fontaine 2017.

[97] Abourezk 2018.

[98] Hammel 2018.

[99] Osaka 2018; IEN 2018b; 350.org 2018.

[100] Aronoff 2017; Grossman 2017; Minnesota Public Radio 2018.

[101] Gardner and Gordon 2018.

[102] Silverstein 2018.

Throughout, the Alliance *as* alliance, and the work members did to build and maintain it, remained central. "The beautiful thing," noted Kleeb, "is that we have farmers and ranchers and climate advocates and tribal nations all at the same table working shoulder to shoulder." "If we didn't do those actions," she added, referring to Ponca corn plantings and Solar XL, "people could have very easily stayed in their silos of only talking about climate or only talking about property rights or only talking about sovereign rights.... those actions built trust among all of us."[103]

In its construction, claims, and strategies, the Cowboy Indian Alliance represents a geographically and historically rooted multi-dimensional politics of connection. It can be understood as addressing each of the relational domains intertwined in the production of climate injustice. The dedication of its members is based in socio-ecological connection, rooted by ties linking people to landscape. Members foreground the embedding of their livelihoods and identities in land and water, highlighting the immediate risks the pipeline poses, while also acknowledging the wider, longer term consequences of climatic change. The Alliance is also based in strong socio-spatial ties between Native groups and whites on the land. Their intentional relationships confront historical and contemporary violence and inequality, and strive to span the differently constituted social spaces of Indigenous and settler nations. Finally, the New CIA's statements and strategies of presence confront the governmental effects of corporate influence over the definition of collective interest at state and national levels. The group's words and actions persistently highlight both particular and broader consequences tied to the pursuit of "collective" interest so defined.

To date, the New CIA's efforts have played a key role in at least delaying the construction of KXL. It has also inspired some reflection about the demographic and values bases of environmental and climate politics. Its wider impacts are less certain. Exactly how central alliance-building remains, as key members of the coalition chart paths in electoral politics and other activist projects, remains to be seen. So far, the group appears to reflect aspects Grossman associates with successful prior Indigenous-settler alliances, including a sense of common place, purpose, and understanding in recognition of conflictual histories.[104] Importantly, that model suggests that "success" be understood not only in terms of effective environmental protection, but also as greater equality and acceptance between participating Native and white communities, leading ultimately toward decolonizing social relations and shared places.

---

[103] Ablow 2017.
[104] Grossman 2017, 273–290.

The last term is critical: Grossman's multi-decadal study of Indigenous-settler alliances highlights efforts to defend "*local* bicultural place[s]"[105]: more successful the more geographically rooted they have been. Certainly for the New CIA, place and its meaning do ground shared rejection of KXL. It so happens, however, that US climate activists also needed grounding and found it in rejecting the pipeline, partly because of the meaning-laden mobilization of the New CIA and other regionally and culturally identified allies. The New CIA is not a "climate justice" organization *per se*, but it brought human faces and stories to climate change politics, articulating them with causes of broader social resonance, in which place plays a key role. The Alliance defends a particular place, and may remain focused there. To the extent, however, that the future of that place and the communities who identify with it or rely on its bounty are subject to fossil fuel-driven climate change and its uneven consequences (as they unavoidably are), that place and the pipeline it may still one day host can also be seen as crucial nodes in a larger network of meaning, connected by the analysis and practices of neighbors who worked to understand, and stand with, one another.

Conclusion

The ECJP and New CIA ground the imperatives of climate justice by articulating the globally extensive ties and uneven geographies of the fossil energy-climate change process with the established and immediate concerns of existing and emergent political communities. This grounding helps mobilize national and subnational publics, and bring the voices of marginalized affected communities into governance fora, through social mobilization as well as established channels of advocacy. In making these links, grounded efforts can also help to legitimate and support transnational coalitions, as Chap. 6 describes.

The New CIA emphasizes culturally resonant actions and novel (if not unprecedented) solidarities, while the ECJP's narratives and analyses leverage already-assembled constituents and advocacy networks. Both link constituents' core concerns—rights in land and cultural identity in the case of the New CIA; racial justice for the NAACP—with threats from fossil-fueled development. Both socialize, historicize, and politicize those threats, focusing "upstream" on disproportionate damage from the production of climate changing fuels, while acknowledging their inevitable and uneven downstream consequences. Whereas the ECJP translates climate impacts in terms

---

[105] Ibid, 277, italics in original.

of civil rights, opening avenues to power through the NAACP's storied advocacy infrastructure, the New CIA and its allies target the physical infrastructure of pipelines as weak links in the systemic reproduction of environmental and social dispossession.

The following chapter revisits these strategies in wider context, together with those of the activists, advocates, institutions, and social settings examined in previous chapters. Taken together, those strategies suggest relational political possibilities whose theoretical and practical implications extend beyond these cases and beyond discussions of "climate justice."

Bibliography

350.org. (2018). *#SolarXL Resisting Keystone XL by Building Clean Energy in the Path of the Pipeline.* Available at: https://350.org/solar-xl/

Ablow, G. (2017). *An Unlikely Alliance Fights Big Oil.* Moyers and Company. Available at: https://billmoyers.com/story/making-change-jane-kleeb-keystone/

Abourezk, K. (2013). *Bold Nebraska, Natives Protest Pipeline, Impact on Land.* Retrieved March 19, 2015, from http://journalstar.com/news/local/bold-nebraska-natives-protest-pipeline-impact-on-land/article_3564bde2-962c-5f57-a547-59b81a74394d.html

Abourezk, K. (2018). *Ponca Tribes Reclaim Ancestral Land Along Trail of Tears in Nebraska.* Indianz.com. Available at: https://www.indianz.com/News/2018/06/12/ponca-tribes-reclaim-ancestral-land-alon.asp

Adler, B. (2015). The Inside Story of How the Keystone Fight Was Won. *Grist.* Available at: https://grist.org/climate-energy/the-inside-story-of-how-the-keystone-fight-was-won/

Anft, M. (2013, September 8). Benjamin Jealous Leaves the NAACP a Far Stronger Place. *The Chronicle of Philanthropy.* Available at: http://philanthropy.com/article/Benjamin-Jealous-Leaves-a/154431/

Aronoff, K. (2017, September 26). The Unlikely Alliance That Could Stop Keystone and Transform the Democratic Party. *In These Times.* Available at: http://inthesetimes.com/features/keystone_nebraska_climate_activism_jane_kleeb.html

Associated Press. (2014, April 13). *Opponents Carve Anti-Pipeline Message into Field.* Retrieved March 19, 2015, from http://bigstory.ap.org/article/opponents-carve-anti-pipeline-message-field

Berg, M. (2007). *The Ticket to Freedom: The NAACP and the Struggle for Black Political Integration.* Gainesville: University Press of Florida.

Black Leadership Forum, The Southern Organizing Committee for Economic and Social Justice, The Georgia Coalition for the Peoples' Agenda, & Clear the Air. (2002, October). *Air of Injustice: African Americans & Power Plant Pollution.* Retrieved from http://www.energyjustice.net/files/coal/Air_of_Injustice.pdf

Bond, P. (2012). Durban's Conference of Polluters, Market Failure and Critic Failure. *Ephemera, 12,* 42.

Booker, B. (2018). The Poor People's Campaign Seeks to Complete Martin Luther King's Final Dream. *All Things Considered*, National Public Radio, May 14. Available at: www. npr.org/2018/05/14/610836891/the-poor-peoples-campaign-seeks-to-complete-martin-luther-king-s-final-dream

Brown, B. (2014, July 21). *"We Will Fight": Keystone XL Pipeline Foes Fear Worst for Water Supply*. Retrieved March 19, 2015, from http://www.nbcnews.com/news/us-news/we-will-fight-keystone-xl-pipeline-foes-fear-worst-water-n150041

Capriccioso, R. (2014, April 29). *10 Images From Closing Day of the Cowboy and Indian Alliance Protest in Washington* [Text]. Retrieved March 19, 2015, from http://indian-countrytodaymedianetwork.com/gallery/photo/10-images-closing-day-cowboy-and-in-dian-alliance-protest-washington-154644

Chappell, B. (2015). *President Obama Rejects Keystone XL Pipeline Plan*. NPR. Available at: https://www.npr.org/sections/thetwo-way/2015/11/06/455007054/president-obama-expected-to-reject-keystone-xl-plan-friday

Chomsky, N. (2013). *Occupy: Reflections on Class War, Rebellion and Solidarity* (2nd ed.). Westfield: Zuccotti Park Press.

Democracy Now. (2017, November 17). Tom Goldtooth: Carbon Trading is "Fraudulent" Scheme to Privatize Air & Forests to Permit Pollution. *Democracy Now*. Available at: https://www.democracynow.org/2017/11/17/tom_goldtooth_carbon_trading_is_fraudu-lent

Earthjustice. (2014). Court Upholds Air Safeguard that Would Prevent Thousands of Deaths. *Earthjustice*, April 15. Available at: https://earthjustice.org/news/press/2014/court-up-holds-air-safeguard-that-would-prevent-thousands-of-deaths

Ecosocialist Horizons. (2011). *Ecosocialist Prefiguration*. Available at: http://ecosocialistho-rizons.com/prefiguration/

Ellison, G. (2014). *New Price Tag for Kalamazoo River Oil Spill Cleanup: Enbridge Says $1.21 Billion*. M Live Michigan. Available at: https://www.mlive.com/news/grand-rap-ids/index.ssf/2014/11/2010_oil_spill_cost_enbridge_1.html

Featherstone, D. (2008). *Resistance, Space and Political Identities: The Making of Counter-Global Networks*. Hoboken: Wiley-Blackwell.

Fontaine, T. (2014, August 1). *Sides Present Health Concerns, Job Loss Worries at EPA Hearings*. Available at: http://triblive.com/news/adminpage/6532254-74/hearings-build-ing-federal

Fontaine, T. (2017, March 27). 'A Perilous Pipeline': Indigenous Groups Line Up Against Keystone XL. *CBC News*. Available at: https://www.cbc.ca/news/indigenous/indigenous-groups-keystonexl-2017-approval-1.4042381

Gardner, T., & Gordon, J. (2018). U.S. to Conduct Additional Keystone XL Pipeline Review. *Reuters*. Available at: https://www.reuters.com/article/us-usa-keystone-pipeline/u-s-to-conduct-additional-keystone-xl-pipeline-review-idUSKCN1NZ2TH

Goluboff, R. L. (2007). *The Lost Promise of Civil Rights*. Cambridge, MA: Harvard University Press.

Great Plains Tar Sands Resistance. (n.d.). *Why Resist Tar Sands?* Retrieved March 19, 2015, from http://gptarsandsresistance.org/why/

Grossman, Z. (2017). *Unlikely Alliances: Native Nations and White Communities Join to Defend Rural Lands*. Seattle: University of Washington Press.

Hammel, P. (2018, June 15). In Possible Roadblock for Keystone XL, Pipeline Opponents Gift Land to Ponca. *Omaha World-Herald*. Available at: https://www.omaha.com/news/

nebraska/in-possible-roadblock-for-keystone-xl-pipeline-opponents-gift-land/article_cbcf6d2b-1862-5d11-b4dd-5ad81d0378ee.html

Hart, G. (2013). Gramsci, Geography, and the Languages of Populism. In E. Michael, H. Gillian, K. Stefan, & L. Alex (Eds.), *Gramsci: Space, Nature, Politics* (pp. 301–320). West Sussex: Wiley-Blackwell.

Harvey, D. (1996). *Justice, Nature and the Geography of Difference*. Hoboken: Wiley-Blackwell.

Hefflinger, M. (2014, April 6). *#NoKXL Crop Art Unveiling*. Retrieved from http://boldnebraska.org/cropart-2/

Herbert, S., Derman, B., & Grobelski, T. (2013). The Regulation of Environmental Space. *Annual Review of Law and Social Science, 9*(1), 227–247.

IEN (Indigenous Environmental Network). (2018a). *Indigenous Environmental Network*. Available at: www.ienearth.org/

IEN (Indigenous Environmental Network). (2018b, June 12). *Tribes, Landowners, and Climate Groups Expand Campaign to Build Solar Inside Keystone XL Pipeline Route*. Available at http://www.ienearth.org/tribes-landowners-and-climate-groups-expand-campaign-to-build-solar-inside-keystone-xl-pipeline-route/

IEN (Indigenous Environmental Network). (n.d.). *Tar Sands*. Retrieved from http://www.ienearth.org/what-we-do/tar-sands/

Jamieson, A., & Vaughan, A. (2017, March 24). Keystone XL: Trump Issues Permit to Begin Construction of Pipeline. *The Guardian*. Available at: https://www.theguardian.com/environment/2017/mar/24/keystone-xl-pipeline-permit-trump-administration

Jealous, B. (2010). Crisis in Cancún. *The Huffington Post*, December 9. Available at: www.huffingtonpost.com/benjamin-todd-jealous/crisis-in-cancun_b_794688.html

Jealous, B. (2013). This Is the Moment for Action on Climate Change. *Post News Group*, July 17. Available at: www.oaklandpost.org/2013/07/17/oped-this-is-the-moment-for-action-on-climate-change/.

Massey, D., & Rustin, M. (2013). Rethinking the Neoliberal World Order. In S. Hall, D. Massey, & M. Rustin (Eds.), *After Neoliberalism: The Kilburn Manifesto* (Vol. 1, pp. 191–221). London: Soundings.

Michigan Department of Environmental Quality. (2018). *Oil Spill News and Updates*. https://www.michigan.gov/deq/0,4561,7-135-3313_56784%2D%2D-,00.html

Minnesota Public Radio. (2018). *Rivers of Oil* (audio recording). Available at: https://www.mprnews.org/topic/rivers-of-oil

Mitchell, T. (2011). *Carbon Democracy: Political Power in the Age of Oil*. London: Verso Books.

Mock, B., & Patterson, J. (2014). Want to Support Clean Energy? Fight for Voting Rights. *Grist*, July 25. Available at: http://grist.org/politics/want-to-support-clean-energy-fight-for-voting-rights/

Moe, K. (2013, November 22). "Cowboys and Indians" Camp Together to Build Alliance Against Keystone XL [Article]. Retrieved March 19, 2015, from http://www.yesmagazine.org/peace-justice/cowboys-and-indians-camp-together-to-build-alliance-against-keystone-xl

Moe, K. (2014a, April 24). Brought Together by Keystone Pipeline Fight, "Cowboys and Indians" Heal Old Wounds [Article]. Retrieved March 19, 2015, from http://www.yesmagazine.org/peace-justice/brought-together-by-pipeline-fight-cowboys-and-indians-heal-old-wounds

Moe, K. (2014b, May 2). *When Cowboys and Indians Unite—Inside the Unlikely Alliance That Is Remaking the Climate Movement*. Retrieved March 19, 2015, from http://wagingnonviolence.org/feature/cowboys-indians-unite-inside-unlikely-alliance-foretells-victory-climate-movement/

Mueller, T. (2012). The People's Climate Summit in Cochabamba: A Tragedy in Three Acts. *Ephemera, 12*, 70.

Mufson, S. (2014, April 18). Obama Administration Postpones Decision on Keystone XL Pipeline. *The Washington Post*. Available at: https://www.washingtonpost.com/business/economy/obama-administration-postpones-decision-on-keystone-xl-pipeline/2014/04/1 8/0c8d9f04-c72a-11e3-8b9a-8e0977a24aeb_story.html?utm_term=.7c8f9793ec7c

Mufson, S., & Eilperin, J. (2017). Trump Seeks to Revive Dakota Access, Keystone XL Oil Pipelines. *The Washington Post*. Available at: https://www.washingtonpost.com/news/energy-environment/wp/2017/01/24/trump-gives-green-light-to-dakota-access-keystone-xl-oil-pipelines/?utm_term=.cff33e1cc15c

NAACP (National Association for the Advancement of Colored People). (2010a). *Climate Justice Initiative Toolkit*. Available at: http://naacp.3cdn.net/112a13293ef36d1c41_o6m6bktqq.pdf

NAACP (National Association for the Advancement of Colored People). (2010b). *Handbook for Advocacy/Programs*. Available at: http://action.naacp.org/page/-/toolkits/Handbook_FINAL.pdf

NAACP (National Association for the Advancement of Colored People). (2011a, July 14). *NAACP Applauds EPA's Cross-State Air Pollution Rule*. Available at: http://www.naacp.org/press/entry/naacp-applauds-epas-cross-state-air-pollution-rule

NAACP (National Association for the Advancement of Colored People). (2011b, August 4). *NAACP Passes Resolution Supporting Strong Clean Air Act*. Available at: https://web.archive.org/web/20140808054436/www.naacp.org/press/entry/naacp-passes-resolution-supporting-strong-clean-air-act

NAACP (National Association for the Advancement of Colored People). (2013). *Just Energy Policies: Reducing Pollution and Creating Jobs*. Available at: http://naacp.3cdn.net/8654c676dbfc968f8f_dk7m6j5v0.pdf

NAACP (National Association for the Advancement of Colored People). (2018a). *Climate Justice Initiative*. Available at: www.naacp.org/programs/entry/climate-justice

NAACP (National Association for the Advancement of Colored People). (2018b). *Resource Organizations*. Available at: www.naacp.org/climate-justice-resources/resource-organizations/

NAACP (National Association for the Advancement of Colored People). (2018c). *The Latest*. Available at: www.naacp.org/latest/?cat=0&topic=47

NAACP (National Association for the Advancement of Colored People). (2018d). *Civil Rights Legislative Report Cards*. Available at: www.naacp.org/report-cards/

NAACP (National Association for the Advancement of Colored People). (n.d.). *NAACP ECJP—YouTube*. Available at: www.youtube.com/user/Katrina2Copenhagen

NAACP (National Association for the Advancement of Colored People), Indigenous Environmental Network, and Little Village Environmental Justice Organization. (2012).

Coal Blooded: Profits Before People. Available at: www.naacp.org/wp-content/uploads/2016/04/CoalBlooded.pdf

National Resources Defense Council. (2013). *NRDC Policy Basics: Tar Sands*. Available at: https://www.nrdc.org/sites/default/files/policy-basics-tar-sands-FS.pdf

National Transportation Safety Board. (2012). *Enbridge Incorporated Hazardous Liquid Pipeline Rupture and Release Marshall, Michigan July 25, 2010*. Available at: https://www.ntsb.gov/investigations/AccidentReports/Reports/PAR1201.pdf

Osaka, S. (2018). How to Disrupt Keystone XL? Solar Panels, Lawsuits, and Ancestral Land. *Grist*. https://grist.org/article/how-to-disrupt-keystone-xl-solar-panels-lawsuits-and-ancestral-land/

Our Power Campaign. (2016). *Our Power Campaign*. Available at: www.ourpowercampaign.org/campaign/

Owe Aku (Bring Back the Way &International Justice Project). (2014, February 6). *Moccasins on the Ground*. Retrieved March 19, 2015, from http://www.oweakuinternational.org/moccasins-on-the-ground.html

Patterson, J. (2009a, November 17). Natural Disasters, Climate Change Uproot Women of Color (2). *Trouthout*. Available at: http://truth-out.org/archive/component/k2/item/86799:natural-disasters-climate-change-uproot-women-of-color

Patterson, J. (2009b, December 21). The NAACP Offers 10 Lessons From Copenhagen Climate Change Conference. *The Root*. Available at: www.theroot.com/10-lessons-from-copenhagen-1790873906

Patterson, J. (2010a). Your Take: Climate Change Is a Civil Rights Issue. *The Root*. Available at: www.theroot.com/your-take-climate-change-is-a-civil-rights-issue-1790879295

Patterson, J. (2010b). *NAACP Convenes HBCUs in the Gulf Region to Discuss Sustainability Research Agenda*. September 29. Available at: www.naacp.org/latest/hbcu-covening-the-research-agenda-on-the-oil-disaster-and-and-sustaina/

Patterson, J. (2010c), *Vote as If Your Life Depends on It: Especially for US, It Does!* November 1. Available at: https://climatejusticeinitiative.wordpress.com/page/4/

Patterson, J. (2011). *Nature's Fury—Chronicling the Devastating Effects of Climate Change in the US South*. May 8. Available at: https://climatejusticeinitiative.wordpress.com/2011/05/08/nature%E2%80%99s-fury%E2%80%94chronicling-the-devastating-effects-of-climate-change-in-the-us-south/

Patterson, J., & Njamnshi, A. (2011). *From the Bronx to Botswana: Making a Climate Change Connection*. July 30. Available at: https://thegrio.com/2011/07/30/from-baltimore-to-botswana-making-the-climate-change-connection/

Poor People's Campaign. (2018). *Demands*. Available at: www.poorpeoplescampaign.org/demands/

Protect the Sacred. (n.d.). *About*. Retrieved March 19, 2015, from http://www.protectthesacred.org/about

Reject and Protect. (2014a). *Call to Action*. Retrieved March 19, 2015, from http://rejectandprotect.org/call-to-action/

Reject and Protect. (2014b). *Camp Schedule*. Retrieved March 19, 2015, from http://rejectandprotect.org/camp-schedule/

Reject and Protect. (2014c). *Press*. Retrieved March 19, 2015, from http://rejectandprotect.org/press/

Reject and Protect. (2014d). *Reject and Protect: Stop Keystone XL*. Retrieved March 19, 2015, from http://rejectandprotect.org

Roland, G. (2014, April 30). *Indigenous Activists Invoke the Sacred as Keystone Pipeline Standoff Continues*. Retrieved March 19, 2015, from http://www.occupy.com/article/indigenous-activists-invoke-sacred-keystone-pipeline-standoff-continues

Routledge, P. (2011). Translocal Climate Justice Solidarities. In J. S. Dryzek, R. B. Norgaard, & D. Schlosberg (Eds.), *The Oxford Handbook of Climate Change and Society* (pp. 384–398). Oxford: Oxford University Press.

Routledge, P., & Cumbers, A. (2009). *Global Justice Networks: Geographies of Transnational Solidarity*. Manchester: Manchester University Press.

Scheingold, S. A. (1974). *The Politics of Rights*. New Haven: Yale University Press.

Shannon, M. J. (2010). *NAACP Hosted an Interagency Briefing on July 29 in New Orleans*. August 17. Available at: www.naacp.org/latest/interagency-briefing/

Shipp, E. R. (2018, April 7). NAACP Poised to Lead Once Again. *The Baltimore Sun*. Available at: www.baltimoresun.com/news/opinion/oped/bs-ed-op-0418-shipp-naacp-leadership-20180417-story.html

Silvern, S. E. (1999). Scales of Justice: Law, American Indian Treaty Rights and the Political Construction of Scale. *Political Geography, 18*(6), 639–668.

Silverstein, K. (2018). *Questionable Economics Threaten the Keystone XL Pipeline—Not Court Rulings*. Forbes. Available at: https://www.forbes.com/sites/kensilverstein/2018/11/11/questionable-economics-threaten-the-keystone-xl-pipeline-not-court-rulings/#443b1251134f

Stephenson, W. (2015). *What We're Fighting for Now Is Each Other: Dispatches from the Front Lines of Climate Justice*. Boston: Beacon Press.

Sullivan, P. (2009). *Lift Every Voice: The NAACP and the Making of the Civil Rights Movement*. New York: New Press.

Tejada, C., & Catlin, B. (2013, May 8). *Indigenous Resistance Grows Strong in Keystone XL Battle*. Retrieved March 19, 2015, from http://wagingnonviolence.org/feature/indigenous-resistance-grows-strong-in-keystone-xl-battle/

Tincher, S. (2014). WV Shows Dim Efforts in Energy Efficiency. *The (West Virginia) State Journal*, July 24.

TransCanada Corporation. (2018). *Keystone XL Pipeline: Route Maps*. Available at http://www.keystone-xl.com/kxl-101/maps/

Wildcat, D. R. (2013). Introduction: Climate Change and Indigenous Peoples of the USA. *Climatic Change, 120*(3), 509–515. https://doi.org/10.1007/s10584-013-0849-6.

Mapping the Politics of Connection

6

Human beings now wield a dominant, lasting influence over the Earth's environmental and climate systems: this is the present era scientists in multiple disciplines have dubbed the Anthropocene. The framing of an Anthropocene, however, tends to undercut recognition of the profoundly uneven human production of climate change and neglect entirely its starkly disproportionate human impacts. Climate change-related responsibility and vulnerability exist in roughly inverted distributions across multiple demographic categories, including age, income, race, and gender, as well as between nation-states. That is, social power tends to correlate directly with emissions and inversely with their negative consequences. Unfortunately, a careful accounting of their sordid history and fractured geographies will not in and of itself arrest the march of greenhouse gas emissions, which continued to accelerate after a brief drop coinciding with the "great recession" of the first decade of the twenty-first century. Political obstacles to halting that march extend fairly predictably from the global and economy-wide character of its causes, so that coordinating effective responses to climate change and even drawing attention to its social inequities have proven extremely difficult. Moreover, it is increasingly clear that while poor and marginalized groups around the world experience the cutting edge of climate impacts (as well as suffering under some response efforts), in high-emitting polities transitioning away from fossil fuel dependence will require the political support and social protection of what have previously been more secure sectors of society, who face risk from energy transition itself, on top of those resulting from decades of neoliberal restructuring (reducing collective resources from taxation, worker's political power, and social safety net programs)

© The Author(s) 2020
B. B. Derman, *Struggles for Climate Justice*,
https://doi.org/10.1007/978-3-030-27965-3_6

and recent austerity measures. Institutionalization, ideological commitment, and ties to state and corporate power all vary among such groups, who could in theory help to both develop and support ecologically effective "just transition" proposals in several polities.[1] In many other settings, of course, high-emissions sector workers facing transition costs are also *among* the poorest and most marginalized.

These are some of the political ecologies within which climate justice advocates have sought means to widen awareness and salience around the uneven and worsening impacts of climate change and the surrounding fossil-fueled political and economic order. The globally linked—but geo-historically situated—conditions of affected groups as well as potential allies and other stakeholders correspond to the diverse range of efforts these activists have pursued, including the ones discussed in the preceding chapters. As the variety in those examples suggests, because of the linkages between climate, culture, economy, security, health, and social justice concerns, the potential points at which compelling connections can be drawn are numerous, as are related, politically resonant knowledges, norms, and practices. Relatedly, the multiplicity of approaches to climate justice indicate the many epistemological and institutional fields in which organizers, advocates, and thinkers must expose gaps and build clear and effective links to create just and effective responses.

Reflecting these conditions, the politics of connection by which climate justice has developed is fundamentally articulatory: re-framing the spatially extensive, more-than-human assemblages of climate injustice to pose "politically key questions in a novel manner."[2] It is also socio-ecological praxis: an intellectual-practical pursuit of liberatory understanding and change.[3] Advocates of climate justice use a combination of analytical, social, and spatial strategies to link the evolving, uneven socio-material realities of climate change with the most promising already-established conceptions and institutions, extant movements and emerging political energies, and—ultimately—to the lived realities and understandings of potential constituencies, giving priority to the historically marginalized among them. The intellectual component of this work involves bringing mainstream climate science together with other forms of knowledge and communication, as the Inuit Circumpolar Council (ICC) did with IQ through elder interviews, for instance, and

---

[1] French Yellow Vest protestors and coal miners in both Poland and the US, for instance, each exerted influence over climate-related policy initiatives in 2018, despite varying in these dimensions.

[2] Escobar 2008, quoted in Featherstone 2011, 139. Featherstone highlights the potential in bringing concepts of articulation and assemblage together.

[3] Following, among others, Hart 2013; Loftus 2013.

as organizers in transnational civil society spaces and National Association for the Advancement of Colored People (NAACP) chapters have done with lay testimonial, feminist political economy, and histories of colonization, enslavement, exploitation, and oppression. In many cases, these and other not narrowly or explicitly "climatic" forms of knowledge are crucial for substantiating or making legible the constitutive connections of climate injustice, but they also shift and broaden the range of social groups whose voices, experiences, perspectives, and prospects emerge and matter. When they valorize and mobilize diverse ways of knowing and communicating, activists expand "environmental politics" socially as well as affectively and epistemologically. Understanding climate change and climate injustice in wider, systemic perspective—as a biogeochemical phenomenon tightly linked with the social relations and material practices of dispossession, privilege, and profit—enables linking them with a wide variety of already-meaningful objects of political force and opportunity. Successful articulation of this kind politicizes climate change, deepening and broadening self-identification by individuals and groups as "stakeholders" in struggles for just and effective responses, and opening avenues for those struggles to unfold.

Reading across the work of the activists and advocates discussed in the preceding chapters, specific attributes of this politics of connection stand out: reflecting, mixing, and developing features of the different domains of relational thinking that shaped those discussions. First, the articulatory analyses and alliances that make climate injustice legible and politically salient often depend upon (and sometimes produce) spatialities and histories that both span the globe and root deeply in the socio-ecological particulars of communities and places. Second, there are notably pragmatic as well as visionary tendencies within these politics, which respond to conjunctural conditions and opportunities as well as the broad sweep of history, and engage a wide range of institutions, norms, and collectives in the face of governmental incapacity, capture, and separation. "Civil society" emerges, in those engagements, as a complex, indeterminate terrain of struggle wherein the governmental may be periodically, if fleetingly, brought face to face with more agonistic forms of the political. Third, and finally, these politics enroll the powers of non-human constituents of climate change and everyday life in creative and critical ways: giving them interpretive context and thereby structured utility in the social process of opposition, resistance, and alternatives. This final chapter explores these themes and indicates some of their wider implications for understanding transnational, emergent, and ecologically oriented movements for social change.

Articulating the Breadth and Depth of Climate (In)justice

Troubling the aggregations of global and species-level framings, by linking environmental changes with human history and social relations, enables organizers to address the concerns of many established transnational social movements—as they do in the open venues convened during international climate negotiations. Climate-related environmental changes and globally networked public-private initiatives assume different meaning for Indigenous, peasant, women's, and left organizations when analyzed in connection with colonialism, patriarchy, and capitalism, just as globally oriented "environmentalism" resonates differently, and more widely, when integrated carefully with a politics of social justice, rather than with that of conservation.

In negotiations as well as outside venues, movement leaders work to frame the more-than-human exploitations, dangers, and possibilities of contemporary earth history in ways that resonate with globally distributed potential constituencies. In socializing the Anthropocene, their analyses run in parallel with Moore's and Haraway's, of the Capitalocene, Plantationocene, and Chthulucene,[4] but on a track joined at more well-signed junctures with pre-existing political blocs, organizations, norms, and energies. Those norms include the salience of human rights, not only as a set of formal legal commitments by nation-states, but as a language of global justice with particular resonance for subaltern claimants and north-south alliances[5]; key provisions in the Framework Convention on Climate Change, like Common But Differentiated Responsibilities (CBDR) and equity; and a view of "civil society" (inside and outside) as a necessary, legitimate counterpart to the narrowly defined national interests that so often dominate international relations and legal forums. Advocates and activists articulate climate justice with these and other prior principles to help frame broadly voiced political speech and to access or create political opportunities. Those concerned with inside-outside linking, in particular, have worked to expand upon existing principles or merge them with climate-specific knowledge, with the aim of additional institutional and societal recognition. This is the case in their work with climate debt, for example, which developed from ecological systems analysis and accounting, and the rights of nature, which fuse Indigenous, anti-capitalist, biocentrist, and environmental legal thought.

---

[4] Moore 2015; Haraway 2015; Haraway et al. 2016.
[5] Santos and Rodríguez-Garavito 2005.

These transnationally oriented socio-ecological and political articulations became the basis of multiple initiatives and networks, with varying resonance in institutions and movements. The linking of institutionally potent norms and principles to climate justice has partially re-shaped discourse in global governance, as recent statements by the United Nations (UN) Office of the High Commissioner of Human Rights (OHCHR) and Special Rapporteur on the Right to Food, and a broad shift toward justice framings among civil society organizations in international settings and others attest. In movement circles, Climate Justice Now!'s (CJN!'s) work in particular has suggested an emerging recognition on the part of far-flung independent groups of world-encompassing, socio-ecological class divisions between the marginalized and privileged, together with the climate-related processes and relations that produced and reproduce that divide. These and other efforts in inter- and transnational settings have offered the points of connection, at which groups around the globe tie their particular concerns with the inclusive, crosscutting project of climate justice.

The "stories-so-far" of affected groups and their allies come together in the climate justice movement's common narratives and spaces of convergence.[6] Activists and advocates have succeeded dramatically in this respect, exploding the spatially and socially exclusive "militant particularisms" that have limited the reach of some efforts for environmental or social protection.[7] They instead build upon less well-documented precedents of transnational subaltern solidarity, like those documented by Featherstone.[8]

Norms and principles vary, however, in their resonance among frontline and potentially allied political communities, as debate during COP17 over the rights of nature suggested (see Chap. 4). Universal proposals risk eliding the particular conditions of socio-ecological dispossession faced by affected communities distributed around the globe. These and other potentially crucial constituencies may already be deeply invested in geo-historically situated emancipatory projects. The issues South African organizers and others raised at University of KwaZulu-Natal (UKZN) illustrated the constitutive, if often implicit, role conceptions of nature-society relatedness play within such projects. These conceptions are often deeply tied to experience, place, and identity, as in the example of thoroughgoing dispossession suffered by Black South Africans under colonialism and Apartheid.

---

[6] Massey 2005; Routledge and Cumbers 2009.

[7] See, for example, the concerns raised around environmental justice activism in Harvey 1996.

[8] These traditions are explored extensively in Featherstone 2008, 2012.

Whereas the global socio-ecological and -spatial processes tying climate change with disparity provide potent grist for critique and solidarity in transnational campaigns and meetings, the specific impact of those mechanisms on places and identity groups necessarily underpin their relevance for many potential constituencies. As the NAACP and New Cowboy Indian Alliance's (CIA) efforts suggest, the analytical and synthetic work of organizers rooted in socio-spatial context is crucially important in rendering the globally extensive, complex relations and mechanisms of climatic injustice legibly connected with the everyday realities and pre-existing issues of affected groups and their allies.[9] As discussed in Chap. 5, those immediate concerns include precarious life in poor, Black neighborhoods adjacent to aging coal-fired power plants, refining centers, and/or rising and warming seas; and the fragile but majestic landscape atop the Ogallala Aquifer, where a history of violence inflects contemporary solidarities with particular gravity.

Clearly such "grounded" articulations differ from those in inter- and transnational settings in some important ways. They can help to mobilize intertwined elements of place, history, identity, and citizenship in the construction of critique, resistance, and counterproposals in a more sustained fashion over time. They can thereby encourage existing constituencies to take on board climate justice-related concerns, and provide pathways for impact by leveraging more well-established social and political infrastructures, when and where those exist. The Environmental and Climate Justice Program (ECJP) does this with the US civil rights tradition and the NAACP's membership, media presence, institutional ties, and capacity for policy and legal advocacy; the New CIA with treaty and property rights, as well as local, state, and federal-level legal and political mechanisms by which they blocked the Keystone XL pipeline (KXL).

Just as in the inter- and transnational examples of the ICC, the Maldives, the Conference of the Parties (COPs), and people's spaces, however, varied tactics of communication and action and multiple points of articulation can be deployed in geographically and socially grounded efforts, like those of the ECJP and the New CIA. Widespread tactics of institutional advocacy and public protest, including spatial strategies like those catalogued by Routledge, are deployed by activists in all these contexts, but so, importantly, are other ways of communicating and acting.[10] To wit, the NAACP's narrative-based outreach with members and allies; Nnimmo Bassey's poem at KlimaForum; the round dances, water ceremonies, and painted teepee of Reject and Protect; caravaner testimonials, La Via Campesina's

---

[9] See also Derman 2013.
[10] Routledge 2017.

Mystica, and wall murals at Cancun's Global Forum; and songs of protest and presence in Durban: each are crucial ways of affectively enlivening and transmitting the articulations of climate justice.

Again, however, the context in which articulations take place also obviously matters for their meaning and resonance. Horses, Stetson hats, and traditional regalia signified the unusual coalition behind the New CIA's anti-pipeline message on the National Mall in Washington, DC, with more relevance for US politicians than they might have had for many delegations in Katowice during COP24, for instance.[11] Similarly, both human rights and the rights of nature vary dramatically in their legal and social relevance across polities and cultures. Still, grounded socio-ecological advocacy and activism offers even more points of articulation in existing movements, norms, and institutional levers than transnational ones. Accordingly, for instance, across different settings advocates are assaying a growing range of legal doctrines, including nuisance and public trust, in legal mobilization for climate justice that is only expanding in its range of theory and jurisdiction.

Spiritual and economic imperatives are also growing areas of CJ articulation and advocacy in multiple political settings. Just as Reject and Protect and the New CIA's alliance-building camps illustrated the centrality of spiritually motivated environmental stewardship for its Indigenous members, faith communities and coalitions are re-orienting or forming anew to advocate for climate justice. An array of campaigns and longstanding or emerging networks active in local communities, international meetings, and at every scale between, illustrates the commitment to stewardship and justice that cuts across many faiths, at least for some practitioners.[12]

Meanwhile (and sometimes in connection with faith-based efforts), campaigns and coalitions for a "just transition" and/or "green jobs" have formed around the world at differing scales, linking the imperatives of energy transition and social justice, giving rise, in several polities, to official plans and policies. These groups and initiatives range widely in their recognition and embrace of systemically linked interests and issues, including economic development, organized labor, corporate influence, community-based energy sovereignty, Indigenous rights, and

---

[11] See author comments in reader debate on Moe 2014.

[12] Many were in evidence in Durban, South Africa, for instance, at a pre-COP stadium event featuring a diverse array of faith community representatives. Pope Francis' encyclical on "Care for Our Common Home" is perhaps the most widely cited religious position on climate change. Francis 2015.

scientifically based mitigation targets.[13] Whereas narrower conceptions of a just transition may focus largely on economic protection for regions and workers dependent on an imperiled coal industry, others encompass more or different social-ecological connections. In the US state of Arizona, for instance, the Black Mesa Water Coalition brought a campaign to close a remnant coal slurry pipeline and gain water rights for Navajo communities together with efforts to revivify traditional livelihoods and facilitate small-scale sustainable enterprise.[14]

That is, many of the strongest threads in a politics of connection necessarily develop out of, expose, and build links between the felt needs of communities and the global, more-than-human processes that concentrate climate change-related privilege and threat within different social groups. These links are likely to be more clear and resonant when couched in terms endemic to communities themselves, and vary accordingly. At the same time, however, the long history and reach of intertwined climatic, energy, and economic systems means that even locally oriented analyses and alliances of climate justice can be spatially and temporally complex, and novel.

Spaces and Times of Connection

In some ways, climate justice resembles other national and global movements. Like activists for other causes, climate justice organizers often work as part of larger networks and alliances. They may shift framings and tactics, when necessary, responding to conditions and opportunities specific to any of a wide range of social and political settings. When possible, they may "jump scale" or "venue shop" to access more auspicious conditions in different administrative settings,[15] and, as they frame or re-frame the imperatives of climate justice, they sometimes help to reconfigure political venues or open new windows of opportunity for their cause.

The politics of climate justice is still relatively new, though, and the climate system, knowledge of it, and institutions focused on it are each evolving rapidly. As activists emerge, join, and respond to these evolutions, they are often compelled to innovate. In doing so, they dispel static and simplistic notions about how space and time become part of the politics of social change—and sometimes operationalize

---

[13] See, for example, the Rosa Luxemburg foundation's collection of just transition-related studies. Rosa Luxemburg Stiftung NYC n.d.

[14] Black Mesa Water Coalition 2015.

[15] See, for example, Cox 1998.

new, more complex and creative ones. As Chap. 4 showed, for instance, network structures and partnerships are crucial to effective climate justice advocacy and alliance-building, particularly in helping to give voice to globally distributed, marginalized groups within trans- and international dialogue at COPs and alternative summits. But these spatially expansive associations are also contingent on the diligent efforts of organizers, and often encompass a variety of frictions. Like the "Global Justice Networks" Routledge and Cumber analyzed, alliances and coordination between groups in these settings can be fraught with uneven power relations, conflict, and uncomfortable compromise.[16] Nor can the nature of interaction or priority among actors or processes associated with different political scales and settings be assumed in climate justice alliances, by reading down (or up) the levels of a spatial or administrative hierarchy. Instead, actors and events in the movement repeatedly suggest shifting relations of co-production among groups, individuals, and events across political sites and scales. For instance, as youth community activists at COPs explained, inspiration for action can come from "above" (in transnational dialogue) or "below" (in the grit of local communities), and pass back and forth between them. As in other causes, places have sometimes served to focus CJ activity or concern, but just as often it is the linkages between them that generate potent new understandings and practices. Together with longstanding and emergent climatic, social, and governmental arrangements, once again, it is the analyses activists produce that drive the spatial and temporal particularities and innovations of climate justice politics.

While mainstream environmental and climate organizations have often framed the stakes of their work in wider geographic terms, local chapters and locally relevant concerns remain the cornerstones of broad-based mobilization strategies in most issue areas. Conditions and concerns in places, accordingly, are a crucial part of grounding the politics of climate justice. Examples of place-based mobilization from the cases examined here include "toxic tours" conducted in neighborhoods surrounding refineries at the port of Durban during COP17; testimonials of frontline resistance, such as those that women of color organizers contributed to the ECJP's online video collection; and the New CIA's landscape-based, culturally defined opposition to KXL. The New CIA's example, perhaps more than any other, resonates with aspects of the model of "site"-based activism suggested by thinkers like Woodward et al., which privileges the generative, independently sufficient character of relations internal to sites or places.[17] Yet even in this example,

---

[16] Routledge and Cumbers 2009.
[17] Woodward et al. 2010; cf. Featherstone 2011.

organizers' insistently place-based analyses and tactics came to link their pipeline opposition, with mutually reinforcing consequences, to the broader efforts of climate organizers; as they were in fact already linked with and informed by experiences from prior "unlikely alliances" elsewhere.

Indeed, the broad resonance achieved by the New CIA, like earlier Indigenous-settler alliances, is likely due in part to what Grossman described as the conscious "interweaving of particularism and universalism" in their efforts.[18] In building its alliance on members' common commitment to site-specific socio-ecological relations, the group deployed practices culled from specific Indigenous spiritual traditions and shared the intergenerational narratives of individual farmers. Those communicated, in particularistic ways, the group's more broadly relatable dedication to forms of autonomy and mutual respect, founded on responsible, harmonious environmental stewardship. For similar reasons—and because of their articulation with climate change and justice concerns—the broader set of recent pipeline protests have become the locus of quite extensive alliances and solidarities, both Indigenous and non, as discussed in Chap. 5 and below.

The ECJP's evocations of disproportionate impact and grassroots resistance accrue their force in part from the particularity and immediacy of personal narratives, but also because of their number and variety, which are legible through the group's systemically oriented analysis. People of color are disproportionately affected by the depredations of climate injustice because at every moment of the fossil energy-climate process these historically marginalized groups bear the brunt of the worst environmental consequences and externalities that process entails. Only through analysis of that process *as* a system—one interlinked through time and space with racialized oppression—can vivid images and narratives of impact in communities be brought together and recognized for their broader sweep and pattern. As the ECJP's statement of solidarity with "sisters and brothers" in many parts of the world,[19] its collaboration with Indigenous Environmental Network (IEN) and Little Village Environmental Justice Organization, and the information-sharing sessions it convened with Pan-African Climate Justice Alliance in Durban suggest, the extent of its practice matches that of its analysis.

All the way "up" to the transnational level, then, such grounded efforts can inspire, reinforce, and legitimate wider solidarities. Seeking to deepen those solidarities or advocate concrete, shared positions in their name may, of course, expose the tensions inevitable when different groups and interests assemble, as explored in

---

[18] Grossman 2017, 289.

[19] NAACP 2018.

Chap. 4. Resolving those tensions is the challenging, time-consuming process organizers seek to facilitate with their intentionally inclusive gathering spaces and epistemologically heterogeneous social practices. In at least one sense, ironically, climate change is on their side in this endeavor. To the extent grounded climate justice initiatives resemble the "militant particularisms" that populate social movement histories, their particularities are necessarily the kind that seek to construct universals worthy of their constituencies' commitment, rather than to attempt withdrawal in enclaved security.[20] This is because bringing specific grievances into relation with climate change means apprehending and interpreting its extensive constitutive ties, as they shape life in places and communities differently around the globe. In this sense, climate justice analyses create analytical purchase for wider solidarity, because they open onto wider understanding.

Conversely, local organizers' participation in and around international negotiations, and their knowledge of processes within them, also enable new analyses and solidarities. One such example is that explained by Isabella Zizi during COP23,[21] in which she and other Richmond, California-based activists linked their local Chevron refinery to a constellation of Indigenous groups affected by oil extraction in Canada and the Amazon, as well as violent forest enclosures in Brazil, which are tied to those same production sites through carbon offset markets like those included in California's climate policy. The result is a transnational network of affected groups, which mirrors key locations in the production and regulatory apparatus that sustains the fossil energy/climate change system.

In temporal terms too, climate justice protests and analyses draw on broad patterns as well as rapidly changing conditions and events. Analyses of carbon colonialism and climate debt, for instance, underscore continuities in the global co-production of social marginalization, uneven economic development, and climate change, which span much of recorded history. On the other hand, Filipino activists quickly altered domestic and international climate policy debates in response to the focusing events of recent major typhoons, and Zizi's connection between groups affected by extraction, refining, and offsetting reflects critical appraisal of still-emerging policy "solutions." Similarly, the New CIA and No Dakota Access Pipeline (NoDAPL) protestors seized on a fleeting political conjuncture (the waning days of an amenable White House and deepening distrust of institutions in the US heartland), helping to enable the first victory for opponents

---

[20] Harvey and Williams 1995; Featherstone 2008.
[21] Democracy Now 2017.

of KXL and prolong the period in which Standing Rock could become a nexus for wider movement building.

The crucial objective, in laying groundwork, tenaciously holding firm, and seizing opportunities, is the same. Activists seek to convert conjunctural eruptions of climate injustice into more propitious lasting conditions: a new balance of forces internationally, an altered popular conception of extraction in relation to national interest, climate-friendly major infrastructure decisions or laws. These politics recognize and exploit the openings Massey indicated in both space and time.[22] A potentially infinite number of connections *are* possible, for the interests that hold the world locked in runaway climate change, as well as the movements aligned against them.

Multiplicity and openness apply as well to the roles taken up by civil society groups involved in climate justice efforts, in ways that can trouble starkly drawn theories of their place in social transformation.

Civil Society, State Power, and the Political Challenge of Climate Injustice

Within the range of groups advocating some form of "climate justice," and more broadly among those allied with marginalized groups in advocating greater responsibility by rich countries, corporations, and elites, a wide range of orientations to official power exist. This heterogeneity begs the question of how to understand the politics of climate justice in relation to institutions, on the one hand, and broad-based and oppositional movements, on the other. These relations are of particular moment, inasmuch as analysts of many different ideological orientations consider climate change to require "transformative" responses and since existing institutions have yet to prove sufficient to the task. As the question was posed in the Introduction, what are advocates of climate justice to do, in light of more than a decade of mobilization and advocacy amidst deepening climate-related crises?

The answer suggested by the foregoing cases is complex, and therefore perhaps less satisfying than some offered by more normative theoretical analyses. As one of the organizers quoted in Chap. 4 put it, amidst debate about the relative potential of engagement inside negotiations versus mobilization outside, "[y]ou got to do everything that's necessary … we got our foot in both areas." The director of a mainstream global non-governmental organization's (NGO) climate program offered a

---

[22] Massey 2005.

strikingly similar assessment in response to my query, during an interview in her Washington, DC, office, about her view of the broad ideological gamut and occasional clashes among civil society groups active there, in Brussels, and around COPs. Each of those organizations, in her view, had its role to play. In different wording and settings and in response to different promptings, both described the need for different forms of engagement. These are not universally accepted or at all comforting views, but they speak to assessments of the problem complex of climate change and climate injustice widely shared by experienced advocates and activists. Many engagements hold promise in relation to one or more of its dimensions, but none is likely to address them all, at least within the temporal window of opportunity science suggests remains.[23] As another advocate, with experience in government and intitutionally-oriented civil society roles put it, hope in "silver bullet" solutions has, for many long time observers of climate politics, run out.

At the same time, a growing number of governments, intergovernmental bodies, and public-private partnerships have taken up the mantle of responding to climate change over the previous decade and a half. An even wider range of institutions arguably hold authority with which they could. Each of these, albeit in widely varying ways, constitute potential targets of advocacy, critique, or partnership for civil society groups. The ICC's petition to the Inter-American Commission on Human Rights (IACHR), 350.org's collaboration in the Maldives' campaign under the auspices of the United Nations Human Rights Council and United Nations Framework Convention on Climate Change (UNFCCC) for a rights-based global climate agreement, the wide range of groups active as observer organizations within the UNFCCC, and appeals by the NAACP and several pipeline-fighting groups to US and Canadian national and subnational governments that were examined in the preceding chapters represent a small fraction of institutionally oriented advocacy focused on climate justice. In these and other examples, as described in Chap. 2, legal mobilization often plays a prominent strategic role. As emerging and existing civil society groups awaken to the imperatives of CJ, and additional institutional forums join the widening constellation of climate governance, the number of potential points of institutional engagement increases.

A primary reason to engage in such official processes is simply that elected governments are, to one degree or another, beholden to the constituents they are charged to represent; their claims to legitimacy depend at some level on their responsiveness. Beyond this, civil society groups identify positive potential in engagement, because many governments, on their own and through their partnership

---

[23] A growing literature on activism within, against, and beyond the state reflects similar analyses. See, for example, Routledge et al. 2018.

with or control of private actors, command resources the deployment of which offers potential for furthering more fair and effective climate change responses. Relatedly, to the extent governments retain mechanisms of responsiveness, and participate in or construct climate-related initiatives that may or may not be conceived in terms of justice at their outset, those mechanisms and initiatives afford opportunities for engagement by organizations and movement representatives. The example of the channel created by Bolivian government leadership (and those of other Bolivarian Alliance for the Americas countries) for a climate justice-oriented state-civil society alliance during the period surrounding the World People's Conference on Climate Change and the Rights of Mother Earth (WPCCC) and COP16 is perhaps the clearest example from the foregoing chapters. Justice advocates take such opportunities, if not out of optimism in government representatives' inclinations as some did in the Bolivian example, then to hold those representatives accountable, with the aim of preventing negative outcomes, as organizers explained in civil society summits described in Chap. 4. As those organizers argued, both action and inaction on the part of governments hold consequences for marginalized and frontline groups. Similarly, engagement inside negotiations creates opportunities for non-governmental observers to support developing countries, as well as groups lacking state representation. Moreover, negotiations (like many governance processes) are both cloistered and highly technical, so that public awareness and assessment of them depends heavily on media coverage, in which governments and other participants therefore have a large stake. In seeking transparency, accountability, wider representation, and a more level playing field for different interests, justice-oriented observers at UN meetings play some of the roles envisioned by laudatory analysts of global civil society: helping to fill the democratic gaps in global governance, and protect the vulnerable.[24] Similarly, in opposing corporate interests and working for the recognition and realization of rights in part through official channels, domestic groups like the NAACP and New CIA follow well-established models of progressive social change.

It should also be abundantly clear, however, that governance, policy, and legal processes have remained woefully inadequate to effectively address anthropogenic climate change and the social inequities with which it is bound up. Some participants in "outside" meetings, others who forego engagement with governmental or alternative summits, and like-minded theorists therefore argue that institutional forms of engagement by civil society groups are also insufficient, if not in fact counterproductive.[25]

---

[24] Kaldor et al. 2012; cf. Derman 2014.
[25] Cf. Wainwright and Mann 2018.

From this perspective, the need is arguably greater for robust alternative political spaces, theories, and tactics involving social movements and communities, including dialogue, broad-based mobilization, civil disobedience, "uncivil" and direct action, and independent, prefigurative efforts by those with opportunities to pursue them.[26] These less institutionally oriented activities can still be encompassed by broader understandings of "civil society."[27] Their social promise includes defining, developing constituencies for, advocating, and demonstrating the possibility of alternatives, and the power of people themselves. These are clear necessities, both within polities where governmental initiatives and frameworks aimed at addressing climate change exist, and where they do not.

Organizers of transnational People's Spaces and demonstrations have attempted to foster inclusive dialogue and genuine debate about climate change, enable the development of solidarity among different groups, and lift up community-led alternatives. They have challenged and would dismantle elite-dominated, "post-political" arrangements of consensus like those that have shaped climate negotiations in exclusionary ways. By gathering active publics outside those official settings (and perhaps eventually creating incursions within them), many aim to dismantle what the speaker quoted in Chap. 4 called the NGO-ization of environmental politics, in which elite civil society actors displace or distort the voices of affected groups and broad publics, failing to adequately represent them, and thereby to challenge or renovate governmental responses. That speaker's call to ally with the "fronts" of struggle encapsulated the need to draw such forces, with little or no presence in official climate politics, into the fray, and perhaps to follow their lead in raising resistance and formulating alternatives. Settings like People's Spaces are hybrid, drawing institutionally engaged participants together with others oriented toward movement building (as well as those focused on linking the two, examined below).

For the most part, the wider variety and possibilities among non-institutionally oriented efforts lie beyond the scope of this volume. Nevertheless, a few that have been peripherally connected with the main cases of this study are suggestive. These include Climate Action Camps; the protests, assemblies, and working group meetings of Occupy Wall Street; and the gatherings organized by Ecosocialist Horizons.

In October 2011, between interviews with NGO and government actors in Washington, DC, I traveled to witness coordinating efforts between the environmental and labor working groups of Occupy Wall Street, then near the apex of its activity. In one exchange, members of the environmental group drafted

---

[26] On the relevance of "uncivil" society mobilization and examples of during COP17, see Ndlovu 2013.

[27] For example, that of Kaldor et al. 2012.

proposed language and visited the much larger labor group's meeting to gain support for a statement to the General Assembly (GA) that would condemn duplicitous claims of economic development by KXL advocates. The exchange was an overture on the part of environment working group members, to ensure that their anti-pipeline message (which was little changed by the encounter) was in fact one of socio-ecological solidarity, and might also achieve greater impact. Environmental working group gatherings were also a meeting point for activists preparing to participate in the large KXL protests that took place the following month at the White House, and others involved in efforts to outlaw fracking in New York State. Internal to the group, activists' priorities of opposition between these different fossil fuel extraction initiatives varied. Still, they collaborated in developing what would become "Climate Justice Day" at Occupy Wall Street, which drew out the links between climate-related disruption, fossil fuel infrastructure development, and financial power.[28]

In July of the same year I attended a Climate Action Camp outside Antwerp, where activists from across Europe gathered for solidarity-building, workshops, direct action, and sustainable collective living. In addition to toxic tours in the region, and various protests and public performances targeting nearby financial, energy, and consumer centers, the camp hosted numerous teach-in style sessions covering genetically modified crops, transcontinental nutrient flows, and permaculture, among other topics. Like Occupy protests, Climate Action Camps laid groundwork for the upwelling of public demonstrations and civil disobedience that erupted in 2018 and 19 (discussed below). Some of those present at the Camp had participated in actions outside the Bella Center in Copenhagen during the 2009 climate negotiations there, but for the most part the focus lay with local, regional, and national opportunities for opposition and responsible ecological practices. Some had vowed to give up air travel, for instance, as part of embodied efforts to support globally respectful ways of life amidst what were, in general, privileged Northern European circumstances.

Some of the praxes in evidence at the Camp exemplified the idea of prefiguration, defined by Joel Kovel and Ecosocialist Horizons as "[t]he potential for the given to contain the lineaments of what is to be," and, in human action, "to try to live and think like the future, right now."[29] As a counterpart to permaculture, non-flying, and other arguably prefigurative practices, Ecosocialist Horizons emphasizes theoretical elaboration and implementation of social relations unyoking communities from fossil-extractive capitalism. The group has organized "convergences" for sharing such ideas and practices, in addition to its education and multi-media publishing projects.

---

[28] Prediger 2011; Sluyter 2011.

[29] Ecosocialist Horizons 2011.

Between Dog and Wolf

Spanning the margins of civil society groups' institutional engagements and "outside" activities in the name of climate justice lie the practices of "linking." Examples in the preceding chapters include efforts to break down the boundary between peoples' spaces and international climate negotiations by accredited groups closely allied with community organizations and social movements, like Friends of the Earth International, Indigenous Environmental Network, and Climate Justice Action. More broadly, linking is a function common to representative and membership-based organizations and networks. Like institutionally oriented and autonomous efforts, this role category too resonates with prominent normative theories of civil society, and takes diverse forms. The ICC's use of images and elder interviews and the centrality of IQ in its legal arguments to the IACHR were discursive forms of linking, bringing the voices and knowledge of its lay constituents into the discourse and forums of international human rights. The NAACP's member-based policy advocacy strategies, which incorporate narrative and testimonial to broaden understanding and resonance of climate injustice and amplify member's voices also embody linking.

These and other groups' linking efforts address crucial factors in the reproduction of climate injustice. In their global survey of non-governmental organizations, Salamon and Sokolowski argue that, while "a major social and economic force in countries throughout the world," civil society is also "not a substitute for government."[30] In the context of climate change, the roles of states should encompass holding private actors (including corporations) accountable, ensuring the protection of vulnerable populations, and realizing the rights of citizens. Even where official mechanisms exist for climate action, however, governments have shown little ability and varying degrees of ambition to shift the energetic basis of the global economy amidst inadequate and often diminishing state capacity, interstate competition, rising corporate capacity and liberty, and closed-door forms of deliberation. Concerned individuals literally cannot enter and participate, of their own will, in securitized intergovernmental debates, much less can broad-based collectives. Connecting the innovations and voices of such individuals and collectives "outside" (at alternative summits, in the streets, and in communities) with governmental goings-on inside is the role carved out by groups and individuals focused on linking at trans- and international levels. That role is self-appointed, but it has for the most part been accepted in the UNFCCC. In ways that generally

---

[30] Salamon and Sokolowski 2004, 56, 30.

conform to liberal political theorists' normative conceptions of civil society, the involvement of critical representative organizations in negotiations has long been tolerated and occasionally lauded, though it is also strictly and sometimes opportunistically limited.[31] It is likely that representative organizations are welcomed in these settings in part because they function instrumentally to legitimate not only intergovernmental proceedings, but also the participation therein of other kinds of organizations (those engaged, for example, through research or implementation roles), since such widely differing groups are often categorized en masse as "civil society."

The ability of groups and individuals to effectively enact linking strategies is, however, contingent on normative conditions within governance arrangements, which can shift. The proliferation of climate governance has created multiple settings for engagement by climate justice advocates, and while these constitute recognizably differing political opportunity structures, they are also often linked by political processes, power relations, and dependencies, as in the tying of national debates with international ones discussed in Chap. 3. It is therefore worth comparing and tracking the inclusion of representative groups across the different settings in which climate justice advocates participate, or perhaps could. If the UNFCCC has traditionally reserved space for representative civil society groups, the European Union (EU) has done so perhaps still more embracingly, though nevertheless within limits and under sustained tension with geo-economic imperatives and corporate input. The US federal government, being in part less concerned to demonstrate democratic legitimacy, has not always reserved the same systematic points of civil society contact, particularly for critical groups.[32] Conceptions of the role of civil society are cultural, learned, and therefore mobile along the vectors that link different governance settings. In traversing those links, they inform participants' understanding of governance itself, as the former US negotiator quoted in Chap. 3 suggested in arguing that representative groups also did not belong in the UNFCCC, where the challenges facing negotiators were primarily "technical" ones. Within the period covered in this study, the UNFCCC and COP host countries have multiple times chosen to enclave more completely the climate negotiations, or barred participation by members of representative groups, sometimes of longstanding observer status.[33]

Linking is also the locus of some discontent surrounding civil society groups and the normative value often attached to their work. Some of these tensions were

---

[31] Arts 1998; Betsill and Corell 2008; Fisher 2010.

[32] Derman 2014; cf. Mahoney 2008.

[33] Fisher 2010; Democracy Now 2010; Rall 2018.

indicated by the challenges linking groups face, discussed in Chap. 4. The translation between lay or movement discourse and the terms of governance, described with respect to the complexity of reducing emissions from deforestation and forest degradation (REDD+) negotiations, suggest the challenge of coming together, as well as the potential losses of meaning and fidelity involved in linking an "outside" with the "inside." The enclaving of climate governance, the specialization of its discourse, and the expectation by some decision makers that civil society speak with "one voice," each exacerbate the potential for distortion, just as they encourage pragmatic impulses toward representation over participation. The notion of representation is itself highly problematic, however. Those who represent are inherently removed, albeit to varying degrees, within networks of power, identity, and discourse from those they represent. In assuming that role, then, representers replace the represented, readily occluding their voices and re-inscribing their status as subaltern.[34] Moreover, even when they work to gather members of frontline communities, in dialogue as well as marches and other actions to exert pressure on decision makers, linking organizers risk objectifying those persons, and perhaps their very well-being.

Nevertheless, most of the cases and analysis in this volume involve linking efforts. This is partly because of the influence they have exerted across the period examined. Linking groups and individuals have played consequential roles in generating influential formulations of "climate justice" as a nexus of political stakes with resonance for a wide range of participating communities and movements, in part because of the centrality of COP and alternative summit settings. Many such formulations were incubated in "outside" venues, like those explored in Chap. 4, and disseminated broadly (sometimes renovated in important ways), as participants in those spaces brought them to confront the conditions and conceptions of situated constituencies. COPs and alternative summits also remained central nodes in the development of climate justice politics in this period, despite the proliferation of other scales of climate governance and social movement articulation, because of specific conjunctural conditions. These included increasing awareness about the unevenness and urgency of climate crises, the window of opportunity that apparently opened around COP15 in 2009 and COP21 in 2015 to renovate the international climate regime, and the fortuitous proximity of multiple COPs (in Copenhagen, Durban, and Paris particularly) to already-activated civil society groups. These factors kept the incubator of outside venues and inside critique warm. As was described, massive political and intellectual energy were focused on

---

[34] Spivak 1988; Nagar and Raju 2003.

the run-up to COP15. When that summit failed to deliver (as participants and observers focused on climate action and social justice adamantly judged it did), the dissemination of climate justice thought took off, together with the downscaling of climate politics, at the same time that critical and justice-oriented perspectives spread among observer organizations active at COPs and among a subset of country parties. These changes are apparent in the increasing deployment of justice discourse at COPs and elsewhere, especially among youth and community-based organizations.[35]

Despite both formal limits and organizing challenges, then, linking groups have played an influential role in defining climate justice positions, particularly during COPs. They will likely continue to do so, by calling governmental actors to account, and by translating and amplifying outside perspectives in inside debates over the definition and implementation of longstanding principles like CBDR and equity, and ascendant ones like loss and damage and just transition.[36] National and subnational examples like those of the NAACP and the New CIA illustrate similar interventions: linking their constituencies' concerns with the exercise of state power, even as they participate in struggles over its role, legitimacy, and extent amidst deepening socio-ecological crises. Of course, the level of their renown in comparison to those of less established groups, and the free media and official publicity that can come with institutional engagement are also factors in these groups' influence.

The complexity and uncertainty in these civil society projects for climate justice have relevance for more broadly framed debates. Swyngedouw has described "governance-beyond-the-state"—arrangements in which market and civil society actors (especially NGOs) provide services once associated with national and local states—as "Janus-faced," in the sense of being riven with contradiction.[37] As such arrangements operationalize shifting logics and technologies of government, often in the name of increasing democracy, they rearticulate state-civil society relations by re-shuffling roles, responsibilities, and powers. They often end, however, by deepening deficits of democracy, and furthering market-led mechanisms of resource allocation. Embedded in this and similar critiques is a wariness of normative theories that hold up "civil society" as among the chief guarantors of democracy and collective good.

Like the figure of civil society that emerges from such analyses, though, the figure of Janus is both different and more complex than common usage—within

---

[35] Farnworth 2018; CJA n.d.

[36] For example, CSO Equity Review 2018; Rosa Luxemburg Stiftung NYC n.d.

[37] Swyngedouw 2005, 2009.

which it primarily connotes duplicity—would suggest.[38] Two-faced effigies of the Roman god were positioned atop doors and gates, over which he was thought to preside. Along with such physical passages, Janus was linked conceptually with transitions, beginnings, endings, conflict, and time itself (sometimes smooth-faced in one of his visages, heavily bearded in the other). Through these connections he seems to have been invoked in a wide range of religious rites as well as before and after battle, and depicted in related civic spaces. Resonant in broader geographical terms as well, his name was often attached to locations on borders. These earlier connections of the Janus persona in fact strongly evoke the liminal position of civil society organizations, and linking groups in particular, in relation to climate governance and the political, as instantiated in "outside" venues, movements, and communities. That position entails presence at the meeting points of difference; orientation both inward and outward, toward the past as well as the future; identification with civic order, social conflict, and transformation all at once.[39]

Of course, godhead metaphors suggest power beyond that of earthly NGOs or social movements, and Swyngedouw's evocation of Janus stood not for civil society in itself, but forms of governance that elevate it. Nor, to be clear, are the aims and impacts of civil society writ large necessarily emancipatory for climate governance. The roster of official UNFCCC observers, for instance, encompasses a wide array of interests, identities, and ideological commitments, as has participation in other settings.[40] Still, the role of civil society in response to contemporary socio-ecological crises is by no means limited to the defense of existing power relations or carrying out the privatizing governmentalities of green neoliberalism.[41] It is hybrid; its aggregate political impact indeterminate in relation to existing hegemonies and possibilities of greater insecurity, social protection, or emancipation.[42] Perhaps a better figuration of its potential, then, is that of a creature between dog and wolf: variously domesticated under, and disruptive of, the dominant social order. In Gramsci's spatial metaphoric terms, civil society writ large is a porous and primary terrain of struggle over the form of a collective socio-ecological future.[43]

---

[38] See, for example, Bonnefoy 1992; Taylor 2000.

[39] Cf. Kaldor, et al. 2012; Ndlovu 2013.

[40] See, for example, Bond's discussion of struggles over critical positioning within African civil society in advance of the 2007 World Social Forum. Bond 2005.

[41] Cf. Chatterton et al. 2013.

[42] These correspond, respectively, to Gramsci's and Fraser's (expanding Polanyi's framework) ways of characterizing the indeterminacy of civil society's political impact. Gramsci 1971; Polanyi 2001; Fraser 2011.

[43] Cf. Gramsci 1971; Buttigieg 1995; Burawoy 2003; Bond 2005.

Efforts within it to ensure or alter social and ecological order must be analyzed empirically as well as critically, with an eye toward their material effects at every stage for the primary actors involved as well as those whose interests they invoke.

Things, Kin, and Social Justice

Fossil fuels convert underground carboniferous deposits into social and political as well as motive power. Those intertwined forces have provided the essential underpinning of capitalist, colonial, imperialist, and geopolitical domination simultaneous with the gathering, wholesale renovation of climate and physical systems they set off. While not reducible to fossil energy's materialities, contemporary social inequality and power relations are therefore inextricably bound up with them. Fossil energy sources and infrastructures, and the changing climate they produce, illustrate especially clearly the profound social and ecological effects that more-than-human mixtures make possible, and the crucial role played by inanimate participants in achieving them. Socially consequential assemblages, objects, and non-human species are ubiquitous, however; they were simply sidelined, for the most part, in modernist accounts of history, politics, and social change.[44] Investigating socio-ecological relatedness involves attending to the work of non-humans within more-than-human mixtures as well as interrogating the social construction and production of the apparently "natural."

Misrecognitions of climate and environmental injustice—like those examined in Chap. 2, or those undone by the ECJP's connective "climate justice 101" trainings—conform to a well-developed theme in critical socio-ecological *re*-tellings of history and power. That theme suggests that where existing power relations are built upon or re-inscribed by more-than-human mixtures, the human actions and social relations that make up those mixtures are readily naturalized, and thereby obscured or accepted.[45] In other narratives centering non-humans objects and species, they exert profound if less obviously political influence.[46] Amidst widening recent concern over climate-related mass extinction and incipient social disintegration, though, analysts have increasingly examined non-humans as potential allies, voices of reason, or exemplars for re-imagining the socio-ecological and, perhaps, the human. In these analyses, often rooted in Indigenous, scientific, and

---

[44]Latour 1993.

[45]See, for example, Castree 2005; Mitchell 2011.

[46]Mitchell 2002; Latour 1993; Robbins 2012.

feminist modes of understanding, non-humans can potentiate more respectful, precautionary, humble, and even liberatory ways of thinking and being.[47]

The preceding chapters examined several moments and mechanisms in which socio-ecological relations and their non-human constituents provide the grounds or resources for resistance, as well as the deepening of exploitation and marginalization. The non-human species and things within socio-ecological process, relations, and conjunctures can play important parts in the politics of climate justice, helping to ground its meanings, build extensive and unlikely alliances, and interpellate new and broader constituencies. When non-humans become what Latour and others have called "actants" in these politics, however, it is through the human analyses that enroll them in laying bare injustice, the need or shape of reconfigured socio-ecologies, or possibilities for resistance. The particular properties of these non-human entities enable their enrollment, without determining it; they must be brought into articulation with emancipatory or liberatory social interests and capacities through corresponding forms of socio-ecological analysis.

The preceding chapters offer several examples in which things and kin (Haraway's term for figures of non-human "nature") get mobilized by advocates and activists, in analyses drawing on diverse forms of knowledge or tactics meant to challenge more-than-human processes and relations of harm, marginalization, or inequality. The ICC's human rights claims depend squarely on recognition of the situated socio-material importance of ice as the literal foundation of Inuit life and culture.[48] Similarly, the Maldives' situated social life in relation to land, fish, salt water, and elevation. Both accounts, and those of other developing countries, UN observers, and outside groups centrally involve emissions accounting, the global atmospheric and ocean circulations, and the molecular behavior, residence time, and industrial-ecology history of greenhouse gases. Alternative futures, and prefigurative pasts and presents framed in terms of *Buen Vivir*, intrinsic value, and the rights of nature assert the kinship and agency of non-human entities, up to and including the planet. The ECJP and the New CIA's analyses, which share in some of those concerns, enfold built infrastructure in more central ways, as coal stations and pipelines join instrumentally in ongoing, historical, and emerging human acts of dispossession against people of color, the Indigenous, and rural settler communities. Recent anti-pipeline and anti-extinction movements illustrate some of the social and material opportunities as well as liabilities of centering the non-human in political and moral campaigns.

---

[47] For example, Shiva 2013; Haraway 2015; Tsing 2015.

[48] Sheila Watt-Cloutier, lead plaintiff in the ICC's petition, elaborated on the importance of ice in her book and lectures on "The right to be cold." Watt-Cloutier 2018.

Flow, Block, and Join

Non-human elements within assemblages matter both materially and politically. Exactly how they do, though, depends quite importantly on social, geographic, and historical contexts, which can shift in consequential ways. In his expansive history of oil and democracy, Timothy Mitchell argues that industrial reliance on coal enabled the emergence of modern democratic politics, lead largely by coal workers. Oil, in contrast, imposed limits on those politics. Those limits extend from several characteristics of oil itself, as well as the geography and history of fossil energy exploitation. In conventional extraction, oil emerges liquid, at high pressure from below ground; capturing it does not require large numbers of workers, such as those that early coal mining operations drew together. Oil can be transported in pipelines and tanker ships, again without the labor of many workers. Seagoing transport networks are more flexible than those of rail or road: oil-bearing ships can be diverted, for instance, to avoid ports temporarily disabled by striking dock workers. Oil also appeared in the first industrialized areas long after coal and its socio-technical infrastructure. Brought from remote, less-industrialized areas (to European centers in particular), oil offered an alternative in terms of organization as well as fuel, the adoption of which enabled states and industries to undercut the political power and democratizing potential of coal miners and movers, who were typically located nearer to population and industrial centers. To control those long, comparatively worker-free networks of international oil transport, multi-national corporations grew up, developing tactics of their own to prevent competition and oversupply, and thereby protect market share and realize astronomical profits. Together with those increasingly powerful corporations, industrial and supplier states took advantage of the material characteristics of oil, and the socio-technical infrastructure surrounding it, to regain the political power coal had helped to give to workers, averting the potential realization of more thoroughgoing, internationalist democratic impulses.[49]

If the history and material properties of oil played key roles in the tamping down of democratic politics during the twentieth century, through, they have since become focal points around which turn tactics and logics of energetic resistance, led not by energy industry workers, but by tribes, landholders, communities, and their allies. These groups have come together and mobilized around analyses exposing how the disproportionate local, regional, and climate-related environmental impacts of fossil energy pipelines enable and symbolize the growing power of globalized capital

---

[49] Mitchell 2011.

and unaccountable governments to preempt citizens' rights and preferences, re-inscribe the dispossessions of settler colonialism, and threaten shared values. Those pipeline-based analyses helped produce a socio-ecological politics of alliance, which cuts across social difference and space in novel and often effective ways, as the examples in Chap. 5 illustrated. How, though, do the materialities of oil and gas play a role in these contemporary politics, as Mitchell argued coal's weight, solidity, and nearby but subterranean location did in earlier, labor-led struggles for democracy? It turns out that pipelines, in particular, are central to the vulnerabilities and points of connection that offer opportunities and motivations for these recent forms of resistance.

Critics note that transporting fossil fuels through pipe involves inevitable risks from leaks and spills. Those eventualities create worrisome vulnerabilities for ecology and health, since, as Enbridge's Kalamazoo River spill dramatized, pipelines often traverse sensitive lands and waters far from the workers tasked with monitoring them, and liquid fuels and flow-enhancing additives readily move into and along with soil, water, and plant and animal life (see Chap. 5). For a number of reasons, though, pipelines are also more vulnerable to physical and political opposition than are other components of liquid fossil energy systems.[50] For one, those long miles of pipe, laid far away from company supervision, offer opponents as much terrain along which to physically occupy infrastructure and thereby incite broader public opposition, as members of the Red Lake Chippewa did, where Enbridge's Line 3 crossed a remote parcel of their land near Bemidji, Minnesota.[51] Elsewhere, activists have physically closed pipeline valves, blocking flow itself.[52]

As the instance of the camp-based action on Red Lake land suggests, however, length and remoteness are only two of the vulnerabilities that afflict webs of fossil energy pipe. To link sources in a continental interior, like Alberta's tar sands or North Dakota's tight oil beds, with distant refineries and export hubs, new or upgraded pipelines must cross a diverse political and physical geography composed of land and water under public, private, and tribal control. Each legal and material conjuncture along those routes may bring new opposition efforts from associated human communities and allies, like those raised by DAPL protestors at Standing

---

[50] See Kenny Bruno's analysis, summarized in Minnesota Public Radio 2018. Ports, too, can become choke points vulnerable to political action, as in the 2015 occupation of Port of Seattle waters by environmental and climate justice "kayaktivists" seeking to prevent arctic drilling by Shell, and efforts to prevent trains reaching that and other Pacific Northwest ports with oil and coal bound for China. Keim and Macalister 2015; Bernton 2016.

[51] Minnesota Public Radio 2018.

[52] Wernick 2018.

Rock and KXL opponents in Nebraska. Particularly during construction, such lines are vulnerable to direct-action strategies, like blocking and locking, by local communities and others gathered in solidarity.

Multiple legal statuses and jurisdictions mean multiple institutional vulnerabilities as well: the more boundaries a pipeline traverses, the more numerous and potentially varied the approval processes its developers must clear. As efforts to block KXL and, more recently, Line 3 illustrate, these processes and the court challenges that may follow offer simultaneous and successive venues in which to contest portions of pipeline routes, across multiple administrative scales. In some cases, challenging a project through administrative and/or direct-action tactics may delay it until other conditions (shifts in wider political support or rulemaking, the development of alternative means of transport or sources of energy) can render a proposed route unprofitable or otherwise unattractive, if not impossible.

Inevitably, perhaps, oil and gas pipelines are vulnerable not only to tactics and politics focused on obstruction, but also to those oriented around connection and flow. In crossing land and water, exacerbating climate change, and indexing the transnational circulation of capital, pipelines materially fuse climate, environmental, and other social movement concerns, facilitating extensive solidarity as well as grounding in community. Whether leaking from or simply moving through pipe, flows of oil and gas have incited novel, re-invigorated, and sometimes unlikely political alliances and identities. Analyses of connection and flow, which draw on history, spiritual tradition, toxicology, and other forms of knowledge, underwrite these impulses. Liquid fuel and its pipelines, it turns out, connect people, issues, things, and places in opposition.

Just as a transnational constellation of opposition emerged from the extraction, refining, and sequestration sites connected by carbon accounting of Shell's California offset, described above, oil and gas flows produce their own socio-spatial politics of resistance. Pipelines themselves trace out potential ties of resistance between communities in their path, as we saw in the example of KXL, which linked first nations and tribal groups, and later drew them together with neighboring white communities. KXL itself incited those communities to join together and then organize to block it, mobilizing around the shared values and experiences they came to recognize.

Unintended flows bring oil and gas into contact with soil, water, flora, and fauna. The mobility of spilled oil highlights the importance and vulnerability of water, as it links humans and non-humans in life-giving flows, repositories, and networks. Water networks and their catchments, too, thereby form the basis of anti-pipeline and related alliances and subjectivities. Thus water protectors, defenders of the sacred, and family farmers have emerged to lead new and re-invigorated movements against pipelines and other manifestations of the fossil energy system, as described in Chap. 5.

Oil and gas workers have largely eschewed and sometimes actively opposed these movements, and industry leadership has launched counter-mobilization efforts, to which a segment of citizens and politicians have responded.[53] Still, many tribes, farmers, and communities who do stand to receive compensation from its development have instead opted to oppose new and upgraded fossil fuel infrastructure. The analyses these pipeline fighters put forth reflect their recognition of the anti-democratic dynamics associated with oil-based development and governance Mitchell chronicled. Their claims are anti-colonial, anti-corporate, and democratic as well as ecologically minded, and these ties draw them closer to the transnational alliances already discussed.

Pipeline opposition efforts sometimes fail, of course, at least in their proximate aims, but activists have made things difficult enough for energy companies to coin new shorthand: by 2018, as they planned new lines developers also worked to devise strategies to undermine or disarm their opponents, who might otherwise "KXL" their projects and profits.[54]

A House on Fire

By 2018, the raw urgency to stop climate change was awakening new activists in the most responsible/response-able countries, and re-orienting the long-dedicated. Like North American pipeline fighters, many identified civil disobedience as crucial components of a social response more "fit for purpose."[55] More fit than thirty years of institutional advocacy, that is, to impel action in the face of stunning neglect by governments and economic elites, and the insidious divorce of policy and public discourse from scientific consensus. The analyses of these movements link ethics and normative political theory with social, cognitive, and climate sciences, and with the perennially youthful energy of rebellion. They aim to force stronger connections in terms of intergenerational and wider social responsibility, policies rooted in science and democracy, and broad mobilization attuned to the lessons of political history and the existential peril all humanity—indeed all species—face in the impending specter of climate breakdown.

---

[53] Minnesota Public Radio 2018.
[54] Minnesota Public Radio 2018.
[55] Extinction Rebellion n.d.

In August 2018, Swedish teen Greta Thunberg sparked mass youth protests when she began sitting on Fridays, alone at first, outside the Swedish Parliament building with a hand-lettered sign that read "school strike for climate." By March 2019, a Global Day of Action by student strikers brought 1.4 million out of their classrooms in 124 counties. They carried placards with slogans such as "why should we go to school if you won't listen to the educated?" and "there is no Planet B."[56] By then Thunberg had addressed the UN climate negotiations in Katowice on behalf of Climate Justice Now! and spoken before the European Parliament. She had chastised decision makers at the World Economic Forum in Davos, saying, "I often hear adults say: 'We need to give the next generation hope,' but I don't want your hope. I want you to panic. I want you to feel the fear I do. Every day. And want you to act. I want you to behave like our house is on fire. Because it is."[57]

Two major climate science reports emerged in late 2018, as Thunberg's allies grew in number. The second volume of the US National Climate Assessment sought to address longstanding worries among Americans and others about the economic impacts of mitigation, quantifying in greater detail than ever before the immense costs of continuing to emit, as well as those "baked in," given the persistence of greenhouse gases already in the atmosphere and their developing system impacts. The following month, the Intergovernmental Panel on Climate Change's (IPCC) "Special Report on Global Warming of 1.5 Degrees C," prepared at the request of climate negotiators, concluded that that aspirational target named in the Paris Agreement would entail impacts already sufficient to threaten numerous ecosystems and human communities, and risks of tipping points previously associated with higher temperatures. As the 2018 COP opened, the group informed negotiators that they had a mere twelve years in which to act, on far more ambitious commitments than parties had offered, if they were to forestall irreversible alterations to multiple earth systems, and widespread social devastation.[58]

Some scientists and observers, however, argued that because of slow summary and reporting processes large scientific bodies like the IPCC were actually retarding the public exposure of more recent and worrying observational data. Worsening the foot-dragging of decision makers, and perhaps reflecting their political resistance, the public focus on major consensus reports was in fact obscuring the true state of knowledge, undercutting the real urgency to act. The organizers of Extinction Rebellion (XR), for instance, highlighted the fact that as

---

[56] Barclay 2019.
[57] Workman 2019.
[58] IPCC 2018.

new observational data become available, they have tended to correspond with or exceed the more extreme possible futures of earlier predictions.

Such scientific truth-telling is a key component in both Thunberg and XR's calls to action. The latter emerged in the UK in 2018 out of prior direct-action campaigns and the Rising Up coordinating collective. XR issued two open letters that year, signed by an international cadre of nearly 200 academic, activist, faith, and political leaders. It then launched a direct-action campaign centered on London, bearing placards with a pictograph of a circled hourglass on brightly colored backgrounds and banners bearing such slogans as "rebel for life," "climate change = genocide," and "hope ends, action begins." The group made three demands: that "government must tell the truth about the climate and wider ecological emergency, reverse inconsistent policies and work alongside the media to communicate with citizens"; that it "enact legally binding policy measures to reduce carbon emissions to net zero by 2025 and to reduce consumption levels"; and that those policies be guided by "a Citizens' Assembly to work with scientists on the basis of the extant evidence and in accordance with the precautionary principle," laying the foundation for "democracy fit for purpose."[59]

XR's analysis and theory of change incorporate climate and extinction science, psychology, ethics, social movement studies, and political history. Organizers scaffolded their argument to groups of potential participants, in a series of steps: paleontological research links four of the five previous mass extinction events to dramatic increases in atmospheric CO2. Climate studies warn of the increasing probability that emissions and policy trajectories are triggering mutually reinforcing positive feedback loops across multiple earth systems that will lead the planet into a dramatically altered "hot Earth" state in the coming decades. Recent human migrations are likely linked already with climate system responses. Nationalism and nativism has taken hold in many countries, raising the specter of fascist resurgence. Together with the coming climate-related collapse of food production, XR argues, these political conditions signal the growing likelihood of mass death, beginning in the most vulnerable communities worldwide, whereby humans may join in the all-but-sure extinction of most other species.

The group encourages facing and experiencing the grief its analysis can inspire as a precursor to willingly bearing a portion of "planetary duty" rooted, for some leading organizers, in spiritual conviction.[60] Current adults have a moral responsibility to act, the

---

[59] The open letters were published by *The Guardian*, which was also the first media outlet to give wide coverage to the groups' efforts. See *The Guardian* 2019.
[60] Cf. Hedges 2015.

group argues, since: "it is unconscionable … that our children and grandchildren should have to bear the terrifying brunt of an unprecedented disaster of our own making." In breaching that responsibility, the UK government (like those of other states, presumably) is guilty of moral and perhaps criminal negligence. In light of foundational political theories spanning a wide ideological spectrum, that neglect legitimates and may even demand rebellion. In this sense, XR argues, to rebel is not a political but a moral act.[61] Given the recalcitrance of political leaders to evidence, then, XR identifies "mass civil disobedience" as the only remaining alternative to avert catastrophe.[62] Finally, the group reads history as suggesting that even a small number of citizens willing to go to jail for a cause can incite significant policy change.

Accordingly, XR mounted several actions in London as well as elsewhere in the UK and Europe during late 2018 and early 2019, dramatizing predictions of mass extinction and death, and making good on its commitment to non-violent law-breaking. Activists have demonstrated in frog headgear, highlighting the place of amphibians at the leading edge of mass extinction, marched in solidarity with student strikers, staged a mock funeral procession to Downing street and Buckingham Palace, held "die-ins" and doused streets in a "blood of our children" demonstration, and blocked bridges and roads. Members have published online videos detailing civil disobedience tactics, glued themselves to buildings, been arrested, and gone to jail. In April of 2019 it launched a coordinated international rebellion focused on major Anglophone and European cities, blocking daily circulation patterns in London during the ten-day action and inspiring British parliamentarians to formally declare a "climate emergency."[63]

Conclusion: Climate Futures and Social Justice

Like several of the campaigns examined in the preceding chapters, pipeline fighters, school strikers, and extinction rebels each deploy critical, creative analyses of the more-than-human. Despite their different emphases, most of those efforts have centered the need for mitigation in industrialized countries, which are, after all, the most responsible for producing climate change in terms of historical, per capita, consumption, and accumulation emissions. Thanks to the levels of economic development they achieved through global

---

[61] Organizers have analyzed the right and duty to rebel in response to climate change in relation to political theories originating with both Locke and Hobbes, for instance.

[62] Taylor 2018.

[63] *The Guardian* 2019.

fossil fuel and human exploitation, they are also arguably the most response-able: exceptionally capable to take action to halt climate change and protect those must vulnerable. The emergence of these movements re-opens old questions, however, around the political implications of analyses framed in terms of infrastructure, urgency, generational continuity, and the survival of species.

As XR's stark summary of climate emergency gained traction through early 2019, it also raised concerns among advocates of climate justice. Critics have noted that nationhood occupied an oddly prominent place in framing XR's purportedly non-political, internationally relevant climate movement, and that marginalized groups remained largely absent from its ranks.[64] Relatedly, XR's advocacy of civil disobedience seems founded in the assumption of liberal states behaving adamantly so: preserving their own civility in the face of illegal actions, and responding constructively to small numbers of principled, dissident citizens regardless of their backgrounds.

As organizers prepared for the "international rebellion" of April 2019, and country chapters framed the movement's demands for their own context, elements within XR began responding to these concerns. South African participants, for instance, called for its national government to "demand climate justice from industrialized countries to reach the budgets required for a just transition to renewable energy sources."[65] The "international solidarity" highlighted in contemporaneous statements by the UK group seemed to signal its growing awareness of the need to avoid the "lifeboat ethics" of population- and security-oriented environmentalisms as well. Into the fall of 2019, and XR's second major protest, both criticism from outside and movement within the group continued. Implicitly responding to established members' missteps on race and immigration, for instance, XR Youth emerged in the UK and beyond, shifting its analyses and partnerships to raise climate justice issues, as XR in the US added a fourth demand devoted entirely to CJ.

The larger point is that to the extent the urgency of averting earth system collapse contributes to globally aggregated, temporally focused analyses, socio-spatial differences in the fate of human groups may more easily be overlooked, or even accepted.[66] Concern for the species writ large, or future generations in particular, may well grow the numbers of the actively concerned,

---

[64] See Trafford's 2019 analysis.
[65] Extinction Rebellion South Africa 2019.
[66] Cf. Harvey 1974.

but they may also elide the stark differences in vulnerability and resources for survival and thriving that meaningfully differentiate the human species in every moment of the deepening crisis, and have done throughout its making. The urgency to halt climate change is real, of course, and increasing daily, but an underlying, uneven socio-ecological geography incontrovertibly variegates the severity of all but its ultimate consequence. As CJN! suggested in Durban, and other analysts have argued since, the future many parents and grandparents in the Global North fear is here now for masses in the South.

Pipeline protests point toward a second dilemma facing climate change and climate justice advocates and the polities and decision makers they work to influence. It is clear that fossil fuels and their despoiling infrastructure must be dismantled and replaced, but the climate crisis has re-opened debate, among scientists and activists alike, over alternatives. In recent years, several have abandoned positions of erstwhile opposition to nuclear energy, sometimes coming round to strongly advocate fission over renewable energy sources. Nuclear advocates' reasons span and often combine engineering, economic, security, and ecological considerations, as well as the time and space required for infrastructure development. The debate over nuclear power is by no means resolved, however, with others arguing strongly against. Their objections include longstanding environmental and security concerns tied to uranium mining and transportation, accidents, waste transportation and storage, weapons development, and terrorism.[67] Many social and climate justice advocates also worry about deepening corporate and state control over communities and their energy supplies tied to the development and maintenance of centralized, securitized, and costly nuclear systems. Similar to activist projects founded on species identity, multi-species commonality, or data-based global abstractions, then, choosing infrastructural allies in the fight to forestall climate breakdown and promote social justice necessarily entails material and political consequences.

Some outcomes cannot be known in advance, of course. The global climate trajectory under current energy and economic regimes is clear, though, as are its sub-global, cruel ironies. This book has argued that those patterns are produced by the socio-material ties of fossil-fueled uneven geographical development, together with the epistemological and political elisions of modernity. These are destructive forms of relationality, which still, in large part, underpin

---

[67] See, for example, KCRW 2019.

the discourses and practices that produce "nature" and society, social space, and political power. Analyses and tactics exposing those relationalities, and constructing morally defensible, safe alternatives underpin the efforts for climate justice examined here. The most thoroughgoing strands of this multifaceted politics of connection reflect the experiences and visions of historically marginalized and frontline communities; the most viable develop extensive alliances, facilitate grounding in communities, and interpellate new constituencies. The more pragmatic among them work for "revolutionary reforms" within existing institutional arrangements[68]; the more visionary for transformation "on the outside," in the streets, and on the land. On the one hand, these efforts provide case studies, highlighting the conditionality and potential within particular articulations. On the other hand, the commonalities among them suggest broad outlines for myriad possible projects, and wider alliances between them, by which we may know climate justice and further it everywhere.

Bibliography

Arts, B. (1998). *The Political Influence of Global NGOs: Case Studies on the Climate and Biodiversity Conventions.* Utrecht: International Books.

Barclay, E. (2019, March 15). Photos: Kids in 123 Countries Went on Strike to Protect the Climate. *Vox.* Retrieved June 10, 2019, from https://www.vox.com/energy-and-environment/2019/3/15/18267156/youth-climate-strike-march-15-photos

Bernton, H. (2016, 15 January). 5 Activists Convicted of Trespass on Everett Tracks in Protest over Oil, Coal Trains. *The Seattle Times.* Retrieved June 10, 2019, from https://www.seattletimes.com/seattle-news/environment/5-activists-convicted-of-trespass-on-everett-tracks-in-protest-over-oil-coal-trains/

Betsill, M. M., & Corell, E. (2008). *Ngo Diplomacy: The Influence of Nongovernmental Organizations in International Environmental Negotiations.* Cambridge, MA: MIT Press.

Black Mesa Water Coalition. (2015). *Our Work.* Black Mesa Water Coalition. Retrieved June 11, 2019, from https://www.blackmesawatercoalition.org

Bond, P. (2005). Gramsci, Polanyi and Impressions from Africa on the Social Forum Phenomenon. *International Journal of Urban and Regional Research, 29*(2), 433–440.

Bonnefoy, Y. (1992). *Roman and European Mythologies.* Chicago: University of Chicago Press.

Burawoy, M. (2003). For a Sociological Marxism: The Complementary Convergence of Antonio Gramsci and Karl Polanyi. *Politics and Society, 31*(2), 193.

Buttigieg, J. A. (1995). Gramsci on Civil Society. *Boundary 2, 22*(3), 1–32. https://doi.org/10.2307/303721.

---

[68] Primrose 2013.

Castree, N. (2005). *Nature*. Abingdon/New York: Routledge.

Chatterton, P., Featherstone, D., & Routledge, P. (2013). Articulating Climate Justice in Copenhagen: Antagonism, the Commons, and Solidarity. *Antipode, 45*(3), 602–620. https://doi.org/10.1111/j.1467-8330.2012.01025.x.

Climate Justice Alliance. (n.d.). *It Takes Roots at COP24*. Climate Justice Alliance. Retrieved June 10, 2019, from https://climatejusticealliance.org/cop24/

Cox, K. R. (1998). Spaces of Dependence, Spaces of Engagement and the Politics of Scale, or: Looking for Local Politics. *Political Geography, 17*(1), 1–23. https://doi.org/10.1016/S0962-6298(97)00048-6.

CSO Equity Review. (2018). *After Paris: Inequality, Fair Shares, and the Climate Emergency*. Manila/London/Cape Town/Washington, et al.: CSO Equity Review Coalition. [civilsocietyreview.org/report2018] https://doi.org/10.6084/m9.figshare.7637669

Democracy Now. (2010, December 8). Youth Activists Protest Exclusion from U.N. Climate Summit in Cancún. *Democracy Now*. Retrieved June 10, 2019, from http://www.democracynow.org/2010/12/8/indigenous_youth_groups_hold_climate_justice

Democracy Now. (2017, November 17). Tom Goldtooth: Carbon Trading is "Fraudulent" Scheme to Privatize Air & Forests to Permit Pollution. *Democracy Now*. Available at: https://www.democracynow.org/2017/11/17/tom_goldtooth_carbon_trading_is_fraudulent

Derman, B. B. (2013). Contesting Climate Injustice During COP17. *South African Journal on Human Rights: Climate Change Justice: Narratives, Rights and the Poor, 29*(1), 170–179.

Derman, B. B. (2014). Climate Governance, Justice, and Transnational Civil Society. *Climate Policy, 14*(1), 23–41.

Ecosocialist Horizons. (2011). *Ecosocialist Prefiguration*. Available at: http://ecosocialisthorizons.com/prefiguration/

Escobar, A. (2008). *Territories of Difference: Place, Movements, Life, Redes*. Durham: Duke University Press.

Extinction Rebellion. (n.d.). *Extinction Rebellion US Demands*. International Extinction Rebellion. Retrieved June 10, 2019, from https://xrebellion.org/xr-us/demands

Extinction Rebellion South Africa. (2019). *Our Demands*. Extinction Rebellion South Africa. Retrieved June 10, 2019, from https://xrebellion.org.za/materialis/our-demands/

Farnworth, E. (2018, December 17). *What Just Happened? 5 Themes from the COP24 Climate Talks in Poland*. World Economic Forum. Retrieved June 10, 2019, from https://www.weforum.org/agenda/2018/12/reflections-on-two-weeks-of-climate-talks/

Featherstone, D. (2008). *Resistance, Space and Political Identities: The Making of Counter-Global Networks*. Hoboken: Wiley-Blackwell.

Featherstone, D. (2011). On Assemblage and Articulation. *Area, 43*(2), 139–142.

Featherstone, D. (2012). *Solidarity: Hidden Histories and Geographies of Internationalism*. London: Zed Books.

Fisher, D. R. (2010). COP-15 in Copenhagen: How the Merging of Movements Left Civil Society Out in the Cold. *Global Environmental Politics, 10*(2), 11–17.

Francis, P. (2015). *Laudato Si: On Care for Our Common Home*. Huntington: Our Sunday Visitor.

Fraser, N. (2011). Marketization, Social Protection, Emancipation: Toward a Neo-Polanyian Conception of Capitalist Crisis. In C. Calhoun & G. Derluiguian (Eds.), *Business as Usual: The Roots of the Global Financial Meltdown* (pp. 137–158). New York: New York University Press.

Gramsci, A. (1971). *Selections from the Prison Notebooks* (Q. Hoare & G. N. Smith, Eds.). New York: International Publishers Co.

Grossman, Z. (2017). *Unlikely Alliances: Native Nations and White Communities Join to Defend Rural Lands*. Seattle: University of Washington Press.

Haraway, D. (2015). Anthropocene, Capitalocene, Plantationocene, Chthulucene: Making Kin. *Environmental Humanities, 6*(1), 159–165.

Haraway, D., Ishikawa, N., Gilbert, S. F., Olwig, K., Tsing, A. L., & Bubandt, N. (2016). Anthropologists Are Talking—About the Anthropocene. *Ethnos, 81*(3), 535–564. https://doi.org/10.1080/00141844.2015.1105838.

Hart, G. (2013). Gramsci, Geography, and the Languages of Populism. In E. Michael, H. Gillian, K. Stefan, & L. Alex (Eds.), *Gramsci: Space, Nature, Politics* (pp. 301–320). West Sussex: Wiley-Blackwell.

Harvey, D. (1974). Population, Resources, and the Ideology of Science. *Economic Geography, 50*(3), 256–277.

Harvey, D. (1996). *Justice, Nature and the Geography of Difference*. Hoboken: Wiley-Blackwell.

Harvey, D., & Williams, R. (1995). Militant Particularism and Global Ambition: The Conceptual Politics of Place, Space, and Environment in the Work of Raymond Williams. *Social Text, 42*, 69–98.

Hedges, C. (2015). *Wages of Rebellion: The Moral Imperative of Revolt*. New York: Nation Books.

(IPCC) Intergovernmental Panel on Climate Change. (2018). *Global Warming of 1.5°C*. Retrieved from https://www.ipcc.ch/sr15/

Kaldor, M., Moore, H. L., Selchow, S., & Murray-Leach, T. (2012). *Global Civil Society 2012: Ten Years of Critical Reflection*. Basingstoke: Palgrave Macmillan.

KCRW. (2019, March 7). Is Nuclear Power the Answer to Climate Change? *To the Point*. Retrieved June 10, 2019, from https://www.kcrw.com/news/shows/to-the-point

Keim, B., & Macalister, T. (2015, June 15). Shell's Arctic Oil Rig Departs Seattle as "Kayaktivists" Warn of Disaster. *The Guardian*. Retrieved from https://www.theguardian.com/us-news/2015/jun/15/seattle-kayak-activists-detained-blocking-shell-arctic-oil-rig

Latour, B. (1993). *We Have Never Been Modern*. Cambridge, MA: Harvard University Press.

Loftus, A. (2013). Gramsci, Nature, and the Philosophy of Praxis. In E. Michael, H. Gillian, K. Stefan, & L. Alex (Eds.), *Gramsci: Space, Nature, Politics* (pp. 178–196). West Sussex: Wiley-Blackwell.

Mahoney, C. (2008). *Brussels Versus the Beltway: Advocacy in the United States and the European Union*. Washington, DC: Georgetown University Press.

Massey, D. B. (2005). *For Space*. London: SAGE.

Minnesota Public Radio. (2018). *Rivers of Oil* (audio recording). Available at: https://www.mprnews.org/topic/rivers-of-oil

Mitchell, T. (2002). *Rule of Experts: Egypt, Techno-Politics, Modernity*. Berkeley: University of California Press.

Mitchell, T. (2011). *Carbon Democracy: Political Power in the Age of Oil*. London: Verso Books.

Moe, K. (2014, May 2). *When Cowboys and Indians Unite—Inside the Unlikely Alliance That Is Remaking the Climate Movement*. Retrieved March 19, 2015, from http://wagingnonviolence.org/feature/cowboys-indians-unite-inside-unlikely-alliance-foretells-victory-climate-movement/

Moore, J. W. (2015). *Capitalism in the Web of Life: Ecology and the Accumulation of Capital*. London: Verso Books.

NAACP (National Association for the Advancement of Colored People). (2018). *Climate Justice Initiative.* Available at: www.naacp.org/programs/entry/climate-justice

Nagar, R., & Raju, S. (2003). Women, NGOs and the Contradictions of Empowerment and Disempowerment: A Conversation. *Antipode, 35*(1), 1–13. https://doi.org/10.1111/1467-8330.00298.

Ndlovu, M. M. (2013). Qwasha! Climate Justice Community Dialogues Compilation. Vol. 1: Voices from the Streets. *South African Journal on Human Rights: Climate Change Justice: Narratives, Rights and the Poor, 29*(1), 180–191.

Polanyi, K. (2001). *The Great Transformation: The Political and Economic Origins of Our Time* (2nd ed.). Boston: Beacon Press.

Prediger, J. (2011, November 2). Green Issues and Greenbacks: Occupy Wall Street Connects the Dots. *Grist.* Retrieved June 10, 2019, from https://grist.org/climate-energy/2011-11-01-green-issues-greenbacks-occupy-wall-street-connects-dots-video/

Primrose, D. (2013). Contesting Capitalism in the Light of the Crisis: A Conversation with David Harvey. *The Journal of Australian Political Economy, 71*, 5.

Rall, K. (2018, December 10). *Poland's Restrictions on Protest Keep Representatives from Civil Society Groups Barred from Participating in COP24.* Business & Human Rights Resource Centre. Available at: https://www.business-humanrights.org/en/polands-restrictions-on-protest-keep-representatives-from-civil-society-groups-barred-from-participating-in-cop24

Robbins, P. (2012). *Political Ecology: A Critical Introduction* (Vol. 20). Chichester: John Wiley & Sons.

Rosa Luxemburg Stiftung NYC. (n.d.). *Introducing the Just Transition Research Collaborative. Rosa Luxemburg Stiftung NYC.* Retrieved June 10, 2019, from http://www.rosalux-nyc.org/introducing-the-just-transition-research-collective/

Routledge, P. (2017). *Space Invaders: Radical Geographies of Protest.* London: Pluto Press.

Routledge, P., & Cumbers, A. (2009). *Global Justice Networks: Geographies of Transnational Solidarity.* Manchester: Manchester University Press.

Routledge, P., Cumbers, A., & Derickson, K. D. (2018). States of Just Transition: Realising Climate Justice Through and Against the State. *Geoforum, 88*, 78–86.

Salamon, L. M., & Sokolowski, S. W. (2004). *Global Civil Society: Dimensions of the Nonprofit Sector.* Baltimore: Johns Hopkins Center for Civil Society Studies.

Santos, B. de S., & Rodríguez-Garavito, C. A. (2005). *Law and Globalization from Below: Towards a Cosmopolitan Legality.* Cambridge: Cambridge University Press.

Shiva, V. (2013). *Making Peace with the Earth.* London: Pluto Press.

Sluyter, M. (2011). *Occupy Wall Street and the Environment: Climate Justice Day.* Human Impacts Institute. Retrieved June 10, 2019, from https://www.humanimpactsinstitute.org/single-post/2011/11/12/Occupy-Wall-Street-and-the-Environment-Climate-Justice-Day

Spivak, G. C. (1988). *Can the Subaltern Speak?* London: Macmillan.

Swyngedouw, E. (2005). Governance Innovation and the Citizen: The Janus Face of Governance-Beyond-the-State. *Urban Studies, 42*(11), 1991–2006.

Swyngedouw, E. (2009). The Antinomies of the Postpolitical City: In Search of a Democratic Politics of Environmental Production. *International Journal of Urban and Regional Research, 33*(3), 601–620.

Taylor, R. (2000). Watching the Skies: Janus, Auspication, and the Shrine in the Roman Forum. *Memoirs of the American Academy in Rome, 45*, 1–40. https://doi.org/10.2307/4238764.

Taylor, M. (2018, October 26). 'We Have a Duty to Act': Hundreds Ready to Go to Jail over Climate Crisis. *The Guardian*. Available at: https://www.theguardian.com/environment/2018/oct/26/we-have-a-duty-to-act-hundreds-ready-to-go-to-jail-over-climate-crisis

The Guardian. (2019). Environment: Extinction Rebellion. *The Guardian*. Available at: https://www.theguardian.com/environment/extinction-rebellion

Trafford, J. (2019, March 29). Against Green Nationalism. *Open Democracy*. Retrieved June 10, 2019, from https://www.opendemocracy.net/en/opendemocracyuk/against-green-nationalism/

Tsing, A. L. (2015). *The Mushroom at the End of the World: On the Possibility of Life in Capitalist Ruins*. Princeton: Princeton University Press.

Wainwright, J., & Mann, G. (2018). *Climate Leviathan: A Political Theory of Our Planetary Future*. London/Brooklyn: Verso.

Watt-Cloutier, S. (2018). *The Right to Be Cold: One Woman's Fight to Protect the Arctic and Save the Planet from Climate Change*. Minneapolis: University of Minnesota Press.

Wernick, A. (2018, 12 May). The "Valve Turners": Activists Faced Jail Time to Briefly Stop the Flow of Canadian Crude Oil. *Public Radio International*. Retrieved June 10, 2019, from https://www.pri.org/stories/2018-05-12/valve-turners-activists-faced-jail-time-briefly-stop-flow-canadian-crude-oil

Woodward, K., Jones, J. P., III, & Marston, S. A. (2010). Of Eagles and Flies: Orientations Toward the Site. *Area, 42*(3), 271–280.

Workman, J. (2019, January 25). "Our House Is on Fire." 16 Year-Old Greta Thunberg Wants Action. *World Economic Forum*. Retrieved June 10, 2019, from https://www.weforum.org/agenda/2019/01/our-house-is-on-fire-16-year-old-greta-thunberg-speaks-truth-to-power/

Bibliography

350.org. (2018). *#SolarXL Resisting Keystone XL by Building Clean Energy in the Path of the Pipeline.* Available at: https://350.org/solar-xl/

Abate, R. S. (2010). Public Nuisance Suits for the Climate Justice Movement: The Right Thing and the Right Time. *Washington Law Review, 85,* 197–252.

Ablow, G. (2017). *An Unlikely Alliance Fights Big Oil.* Moyers and Company. Available at: https://billmoyers.com/story/making-change-jane-kleeb-keystone/

Abourezk, K. (2013). *Bold Nebraska, Natives Protest Pipeline, Impact on Land.* Retrieved March 19, 2015, from http://journalstar.com/news/local/bold-nebraska-natives-protest-pipeline-impact-on-land/article_3564bde2-962c-5f57-a547-59b81a74394d.html

Abourezk, K. (2018). *Ponca Tribes Reclaim Ancestral Land Along Trail of Tears in Nebraska.* Indianz.com. Available at: https://www.indianz.com/News/2018/06/12/ponca-tribes-reclaim-ancestral-land-alon.asp

ActionAid. (2009, December). *Rich Countries' "Climate Debt" and How They Can Repay It.* Retrieved from https://www.actionaid.org.uk/sites/default/files/doc_lib/updated_climate_debt_briefing_december_200.pdf

ActionAid. (2011). *Fiddling with Soil Carbon Markets While Africa Burns....* Retrieved March 17, 2015, from http://www.actionaidusa.org/shared/fiddling-soil-carbon-markets-while-africa-burns-0

ActionAid, & 16 others. (2010a, November). *Climate Justice Briefs #1: Climate Debt.* Retrieved from http://www.ips-dc.org/wp-content/uploads/2010/12/1-Climate-debt.pdf

ActionAid, & 16 others. (2010b, November). *Climate Justice Briefs #12: Human Rights and Climate Justice.* Retrieved from http://www.whatnext.org/resources/Publications/Climate-justice-briefs_full-setA4.pdf

Adger, W. N., Paavola, J., Huq, S., & Mace, M. J. (2006). *Toward Justice in Adaptation to Climate Change.* MIT Press. Retrieved from https://books.google.com/books?hl=en&lr=&id=bIn5W04D3dIC&oi=fnd&pg=PA1&dq=climate+justice&ots=HfBuHEXD0P&sig=qA5tL5ofmZc9inPsltRk3Bj2Gf8

© The Author(s) 2020
B. B. Derman, *Struggles for Climate Justice,*
https://doi.org/10.1007/978-3-030-27965-3

Adler, B. (2015). The Inside Story of How the Keystone Fight Was Won. *Grist*. Available at: https://grist.org/climate-energy/the-inside-story-of-how-the-keystone-fight-was-won/

Agamben, G. (1998). *Homo Sacer*. Redwood City: Stanford University Press.

Agnew, J. (1994). The Territorial Trap: The Geographical Assumptions of International Relations Theory. *Review of International Political Economy, 1*(1), 53–80.

Anderson, K., & Bows, A. (2011). Beyond "Dangerous" Climate Change: Emission Scenarios for a New World. *Philosophical Transactions of the Royal Society A: Mathematical, Physical and Engineering Sciences, 369*(1934), 20–44.

Anft, M. (2013, September 8). Benjamin Jealous Leaves the NAACP a Far Stronger Place. *The Chronicle of Philanthropy*. Available at: http://philanthropy.com/article/Benjamin-Jealous-Leaves-a/154431/

APRODEV. (n.d.). *Climate Change*. Retrieved June 26, 2013, from http://www.aprodev.eu/index.php?option=com_content&view=section&id=13&Itemid=11&lang=en

Arctic Council, Assessment, A. C. I., Monitoring, A., Committee, I. A. S., & others. (2005). *Arctic Climate Impact Assessment*. New York: Cambridge University Press.

Aronoff, K. (2017, September 26). The Unlikely Alliance That Could Stop Keystone and Transform the Democratic Party. *In These Times*. Available at: http://inthesetimes.com/features/keystone_nebraska_climate_activism_jane_kleeb.html

Arts, B. (1998). *The Political Influence of Global NGOs: Case Studies on the Climate and Biodiversity Conventions*. Utrecht: International Books.

Associated Press. (2014, April 13). *Opponents Carve Anti-Pipeline Message into Field*. Retrieved March 19, 2015, from http://bigstory.ap.org/article/opponents-carve-anti-pipeline-message-field

Atapattu, S. (2015). *Human Rights Approaches to Climate Change: Challenges and Opportunities*. London: Routledge.

Averill, M. (2010). Getting into Court: Standing, Political Questions, and Climate Tort Claims. *Review of European Community & International Environmental Law, 19*(1), 122–126.

Babbie, E. R. (2007). *The Practice of Social Research*. Belmont: Wadsworth Publishing Co.

Baer, P., Athanasiou, T., Kartha, S., & Kemp-Benedict, E. (2009). Greenhouse Development Rights: A Proposal for a Fair Global Climate Treaty. *Ethics, Place & Environment, 12*(3), 267–281.

Bakker, K. (2005). Neoliberalizing Nature? Market Environmentalism in Water Supply in England and Wales. *Annals of the Association of American Geographers, 95*(3), 542–565.

Banktrack. (2011). BankTrack.org—news—Bankrolling Climate Change. Retrieved March 5, 2013, from http://www.banktrack.org/show/news/bankrolling_climate_change

Barclay, E. (2019, March 15). Photos: Kids in 123 Countries Went on Strike to Protect the Climate. *Vox*. Retrieved June 10, 2019, from https://www.vox.com/energy-and-environment/2019/3/15/18267156/youth-climate-strike-march-15-photos

Barry, A. (2013). *Material Politics: Disputes Along the Pipeline*. Oxford: John Wiley & Sons.

BBC News. (2009, December 19). Copenhagen Deal Reaction in Quotes. *BBC*. Retrieved from http://news.bbc.co.uk/2/hi/science/nature/8421910.stm

Beckett, K., & Herbert, S. K. (2009). *Banished: The New Social Control in Urban America*. New York: Oxford University Press US.

Bell, D. (2011). Does Anthropogenic Climate Change Violate Human Rights? *Critical Review of International Social and Political Philosophy, 14*(2), 99–124. https://doi.org/10.1080/13698230.2011.529703.

Bell, D. (2013). Climate Change and Human Rights. *WCC Wiley Interdisciplinary Reviews: Climate Change, 4*(3), 159–170.

Bell, D. A., Jr. (1980). Brown v. Board of Education and the Interest-Convergence Dilemma. *Harvard Law Review, 93*(3), 518–533.

Benford, R. D., & Snow, D. A. (2000). Framing Processes and Social Movements: An Overview and Assessment. *Annual Review of Sociology, 26*, 611–639.

Bennett, J. (2010). *Vibrant Matter: A Political Ecology of Things.* Durham: Duke University Press.

Berg, M. (2007). *The Ticket to Freedom: The NAACP and the Struggle for Black Political Integration.* Gainesville: University Press of Florida.

Bergmann, L. (2011). *Presentation to the Department of Geography of the University of Washington.* Seattle: University of Washington.

Bergmann, L. (2013). Bound by Chains of Carbon: Ecological–Economic Geographies of Globalization. *Annals of the Association of American Geographers, 103*(6), 1348–1370.

Berkowitz, S. (1997). Analyzing Quantitative Data. In J. Frechtling & L. Sharp Westat (Eds.), *User-Friendly Handbook for Mixed Method Evaluations.* National Science Foundation Division of Research, Evaluation and Communication. Retrieved from http://www.nsf.gov/pubs/1997/nsf97153/chap_4.htm

Bernauer, T. (2013). Climate Change Politics. *Annual Review of Political Science, 16*(1), 130301143509009.

Bernton, H. (2016, 15 January). 5 Activists Convicted of Trespass on Everett Tracks in Protest over Oil, Coal Trains. *The Seattle Times.* Retrieved June 10, 2019, from https://www.seattletimes.com/seattle-news/environment/5-activists-convicted-of-trespass-on-everett-tracks-in-protest-over-oil-coal-trains/

Betsill, M. M., & Bulkeley, H. (2004). Transnational Networks and Global Environmental Governance: The Cities for Climate Protection Program. *International Studies Quarterly, 48*(2), 471–493. https://doi.org/10.1111/j.0020-8833.2004.00310.x.

Betsill, M. M., & Corell, E. (2008). *Ngo Diplomacy: The Influence of Nongovernmental Organizations in International Environmental Negotiations.* Cambridge, MA: MIT Press.

Betsill, M., & Rabe, B. (2009). Climate Change and Multilevel Governance: The Evolving State and Local Roles. In D. A. Mazmanian & M. E. Kraft (Eds.), *Toward Sustainable Communities: Transition and Transformations in Environmental Policy* (pp. 201–225). Cambridge, MA: MIT Press.

Bix, B. H. (2005). Law as an Autonomous Discipline. In P. Cane & M. Tushnet (Eds.), *The Oxford Handbook of Legal Studies* (pp. 975–987). Oxford/New York: Oxford University Press.

Black Leadership Forum, The Southern Organizing Committee for Economic and Social Justice, The Georgia Coalition for the Peoples' Agenda, & Clear the Air. (2002, October). *Air of Injustice: African Americans & Power Plant Pollution.* Retrieved from http://www.energyjustice.net/files/coal/Air_of_Injustice.pdf

Black Mesa Water Coalition. (2015). *Our Work.* Black Mesa Water Coalition. Retrieved June 11, 2019, from https://www.blackmesawatercoalition.org

Blaikie, P., & Brookfield, H. (1987). *Land Degradation and Society.* London: Routledge Kegan & Paul.

Blaut, J. M. (1993). *The Colonizer's Model of the World: Geographical Diffusionism and Eurocentric History.* New York/London: Guilford Press.

Blomley, N. (1994). *Law, Space, and the Geographies of Power*. New York: Guilford.
Blomley, N. (2008). Simplification Is Complicated: Property, Nature, and the Rivers of Law. *Environment and Planning A, 40*(8), 1825–1842.
Blomley, N., Delaney, D., & Ford, R. (Eds.). (2001). *The Legal Geographies Reader: Law, Power and Space*. Hoboken: Wiley.
Bodansky, D. (2010a). Climate Change and Human Rights: Unpacking the Issues. *Georgia Journal of International and Comparative Law, 38*, 511–524.
Bodansky, D. (2010b). The Copenhagen Climate Change Conference: A Postmortem. *American Journal of International Law, 104*(2), 230–240.
Bold Nebraska. (2015). *Bold Nebraska*. Retrieved March 19, 2015, from http://ne.pnstate. org/
Bond, P. (2005). Gramsci, Polanyi and Impressions from Africa on the Social Forum Phenomenon. *International Journal of Urban and Regional Research, 29*(2), 433–440.
Bond, P. (2010). Climate Justice Politics Across Space and Scale. *Human Geography, 3*(2), 49–62.
Bond, P. (2011). *The Politics of Climate Justice*. London: Verso and Pietermaritzburg: University of KwaZulu-Natal Press.
Bond, P. (2012). Durban's Conference of Polluters, Market Failure and Critic Failure. *Ephemera, 12*, 42.
Bond, P., & Dorsey, M. K. (2010). Anatomies of Environmental Knowledge & Resistance: Diverse Climate Justice Movements and Waning Eco-Neoliberalism. *Journal of Australian Political Economy, 66*, 286–316.
Bonnefoy, Y. (1992). *Roman and European Mythologies*. Chicago: University of Chicago Press.
Booker, B. (2018). The Poor People's Campaign Seeks to Complete Martin Luther King's Final Dream. *All Things Considered*, National Public Radio, May 14. Available at: www. npr.org/2018/05/14/610836891/the-poor-peoples-campaign-seeks-to-complete-martin-luther-king-s-final-dream
Bote, T. (2011, November 30). *Rural Women Farmers Speak Out on Climate Change*. Retrieved March 17, 2015, from http://www.actionaidusa.org/zimbabwe/2011/11/rural-women-farmers-speak-out-climate-change
Bradbury, J., & Tompkins, C. F. (2013). *New Report Connects 2012 Extreme Weather Events to Human-Caused Climate Change* [online]. World Resources Institute. Available from: http://www.wri.org/blog/2013/09/new-report-connects-2012-extreme-weather-events-human-caused-climate-change
Braverman, I., et al. (Eds.). (2014). *The Expanding Spaces of Law: A Timely Legal Geography*. Redwood City: Stanford University Press.
Bridge, G., & Perreault, T. (2009). Environmental Governance. In N. Castreeessor, D. Demerittessor, D. L. Director, & B. R. Chair (Eds.), *A Companion to Environmental Geography* (pp. 475–497). Oxford: Wiley-Blackwell.
Broder, J. M. (2009, December 11). U.S. Climate Envoy's Good Cop, Bad Cop Roles. *The New York Times*. Retrieved from http://www.nytimes.com/2009/12/11/science/earth/11stern.html
Broder, J. M. (2011, December 10). In Glare of Climate Talks, Taking on Too Great a Task. *The New York Times*. Retrieved from http://www.nytimes.com/2011/12/11/science/earth/climate-change-expands-far-beyond-an-environmental-issue.html

Broder, J. M. (2012, November 26). U.N. Climate Talks Promise Little Drama. *The New York Times*. Retrieved from http://www.nytimes.com/2012/11/27/business/energy-environment/un-climate-talks-promise-little-drama.html

Brown, B. (2014, July 21). *"We Will Fight": Keystone XL Pipeline Foes Fear Worst for Water Supply*. Retrieved March 19, 2015, from http://www.nbcnews.com/news/us-news/we-will-fight-keystone-xl-pipeline-foes-fear-worst-water-n150041

Brown, M., & Knopp, L. (2008). Queering the Map: The Productive Tensions of Colliding Epistemologies. *Annals of the Association of American Geographers, 98*(1), 40–58.

Building Bridges Collective. (2010). *Space for Movement: Reflections from Bolivia on Climate Justice, Social Movements and the State*. Retrieved from http://www.climatecollective.org/media/uploads/resources/space_for_movement1.pdf

Bulkeley, H. (2005). Reconfiguring Environmental Governance: Towards a Politics of Scales and Networks. *Political Geography, 24*(8), 875–902.

Bulkeley, H., Carmin, J., Castán Broto, V., Edwards, G. A. S., & Fuller, S. (2013). Climate Justice and Global Cities: Mapping the Emerging Discourses. *Global Environmental Change, 23*(5), 914–925. https://doi.org/10.1016/j.gloenvcha.2013.05.010.

Bulkeley, H., Edwards, G. A., & Fuller, S. (2014). Contesting Climate Justice in the City: Examining Politics and Practice in Urban Climate Change Experiments. *Global Environmental Change, 25*, 31–40.

Bullard, R. D. (2008). *Dumping in Dixie: Race, Class, and Environmental Quality* (3rd ed.). Boulder: Westview Press.

Bumpus, A. G., & Liverman, D. M. (2008). Accumulation by Decarbonization and the Governance of Carbon Offsets. *Economic Geography, 84*(2), 127–155.

Burawoy, M. (2003). For a Sociological Marxism: The Complementary Convergence of Antonio Gramsci and Karl Polanyi. *Politics and Society, 31*(2), 193.

Burger, M., & Wentz, J. A. (2015). *Climate Change and Human Rights* [online]. Columbia University Academic Commons. Available from: https://doi.org/10.7916/D8PG1RRD.

Büscher, B., & Arsel, M. (2012). Nature™ Inc.: Changes and Continuities in Neoliberal Conservation and Market-Based Environmental Policy. *Development and Change, 43*(1), 53–78.

Butler, J. (2011, September). *Bodies in Alliance and the Politics of the Street*. Presented at the The State of Things, organized by the Office for Contemporary Art Norway, Venice. Retrieved from http://suebellyank.com/wp-content/uploads/2011/11/ola-reader-full.pdf

Buttigieg, J. A. (1995). Gramsci on Civil Society. *Boundary 2, 22*(3), 1–32. https://doi.org/10.2307/303721.

C40 Cities Climate Leadership Group. (2015). *History of the C40*. Retrieved February 20, 2015, from http://www.c40.org/history

Cameron, E., & Bevins, W. (2012, December 14). *What Is Equity in the Context of Climate Negotiations?* World Resources Institute. Retrieved from http://www.wri.org/blog/2012/12/what-equity-context-climate-negotiations

CAN (Climate Action Network International). (n.d.). *Tag: Gigatonne Gap*. CAN International. Retrieved February 21, 2015, from http://www.climatenetwork.org/category/tags/gigatonne-gap

Canadian Youth Delegation, joined by 137 delegates and 21 observer organizations. (2012). *Fossil Foul: Letter to Christiana Figueres*. Retrieved March 4, 2013, from http://fossil-foul.tumblr.com/

Capriccioso, R. (2014, April 29). *10 Images From Closing Day of the Cowboy and Indian Alliance Protest in Washington* [Text]. Retrieved March 19, 2015, from http://indiancountrytodaymedianetwork.com/gallery/photo/10-images-closing-day-cowboy-and-indian-alliance-protest-washington-154644

Carrington, D. (2010, December 3). *WikiLeaks Cables Reveal How US Manipulated Climate Accord.* Retrieved February 20, 2015, from http://www.theguardian.com/environment/2010/dec/03/wikileaks-us-manipulated-climate-accord

Castree, N. (2005). *Nature*. Abingdon/New York: Routledge.

Castree, N. (2014). The Anthropocene and Geography I: The Back Story. *Geography Compass, 8*(7), 436–449. https://doi.org/10.1111/gec3.12141.

CCAP (Center for Clean Air Policy). (n.d.). *CCAP at Doha COP18*. Center for Clean Air Policy. Retrieved July 2, 2013, from http://ccap.org/programs/ccap-at-doha-cop18/

CDM Watch. (2011, December 3). *Watch This! Progress and Gossip About Carbon Markets at COP 17.* Retrieved from http://carbonmarketwatch.org/wp-content/uploads/2011/12/watch-this-311.pdf

Centre for Civil Society. (n.d.-a). *Home—Events & Action*. Retrieved March 17, 2015, from http://ccs.ukzn.ac.za/default.asp?2,68

Centre for Civil Society. (n.d.-b). *Wolpe Lectures and Reviews 2011*. Retrieved March 16, 2015, from http://ccs.ukzn.ac.za/default.asp?11,22,5,2668#Rights%20of%20Nature

Chakrabarty, D. (2009). The Climate of History: Four Theses. *Critical Inquiry, 35*(2), 197–222. https://doi.org/10.1086/596640.

Chan, K. M., et al. (2016). Opinion: Why Protect Nature? Rethinking Values and the Environment. *Proceedings of the National Academy of Sciences, 113*(6), 1462–1465.

Chapman, M. (2010). Climate Change and the Regional Human Rights Systems. *Sustainable Development Law & Policy, 10*(2), 37–38.

Chappell, B. (2015). *President Obama Rejects Keystone XL Pipeline Plan*. NPR. Available at: https://www.npr.org/sections/thetwo-way/2015/11/06/455007054/president-obama-expected-to-reject-keystone-xl-plan-friday

Charmaz, K. (2004). Grounded Theory. In S. N. Hesse-Biber & P. Leavy (Eds.), *Approaches to Qualitative Research: A Reader on Theory and Practice*. New York: Oxford University Press.

Chatterton, P., Featherstone, D., & Routledge, P. (2013). Articulating Climate Justice in Copenhagen: Antagonism, the Commons, and Solidarity. *Antipode, 45*(3), 602–620. https://doi.org/10.1111/j.1467-8330.2012.01025.x.

Cheung, William, W. L., Vicky W. Y. Lam, Jorge L. Sarmiento, Kelly Kearney, R. E. G., Watson, Dirk Zeller, & Daniel Pauly. (2010). Large-Scale Redistribution of Maximum Fisheries Catch Potential in the Global Ocean Under Climate Change. *Global Change Biology, 16*(1), 24–35.

Cheung, William, W. L., John Dunne, Jorge L. Sarmiento, & Daniel Pauly. (2011). Integrating Ecophysiology and Plankton Dynamics into Projected Maximum Fisheries Catch Potential under Climate Change in the Northeast Atlantic. *ICES Journal of Marine Science, 68*(6), 1008–1018.

Cheung, William, W. L., Jorge L. Sarmiento, John Dunne, Thomas L. Frölicher, Vicky, W. Y. Lam, M. L., Deng Palomares, Reg Watson, & Daniel Pauly. (2013). Shrinking of Fishes Exacerbates Impacts of Global Ocean Changes on Marine Ecosystems. *Nature Climate Change, 3*(3), 254–258.

Chomsky, N. (2013). *Occupy: Reflections on Class War, Rebellion and Solidarity* (2nd ed.). Westfield: Zuccotti Park Press.

CIDSE. (2011). *Climate Justice*. Retrieved June 26, 2013, from http://www.cidse.org/content/sectors/climate-justice/climate-justice.html

CIEL (Center for International Environmental Law). (2011). *Analysis of Human Rights Language in the Cancun Agreements (UNFCCC 16th Session of the Conference of the Parties)*.

Climate Camp. (n.d.). *Climatecamp—YouTube*. Retrieved February 26, 2015, from https://www.youtube.com/user/climatecamp

Climate Connections. (n.d.). *Jacqui Patterson*. Climate Connections. Retrieved March 19, 2015, from http://climate-connections.org/tag/jacqui-patterson/

Climate Impacts Group. (n.d.). *Welcome*. Retrieved March 15, 2015, from http://cses.washington.edu/cig/

Climate Justice Alliance. (n.d.). *It Takes Roots at COP24*. Climate Justice Alliance. Retrieved June 10, 2019, from https://climatejusticealliance.org/cop24/

Climate Justice Now! (2007). *Founding Statement*. Center for Civil Society. Available at: http://ccs.ukzn.ac.za/default.asp?4,80,5,2381

Climate Justice Now! (2008). *Climate Justice Now! Statement*. Carbon Trade Watch. Available at: http://www.carbontradewatch.org/index.php?option=com_content&task=view&id=227&Itemid=95

Climate Justice Now! (2010). *CJN! Network Members (as at November 2010)*. Available at: https://web.archive.org/web/20150901040407/http://www.climate-justice-now.org:80/category/climate-justice-movement/cjn-members/

Climate Justice Now! (2011). *2011 COP17 Succumbs to Climate Apartheid!* Retrieved from http://www.climate-justice-now.org/2011-cop17-succumbs-to-climate-apartheid-antidote-is-cochabamba-peoples%E2%80%99-agreement/

Climate Justice Now! (@CJNatCOP). (n.d.). *Twitter Feed*. Retrieved March 14, 2015, from https://twitter.com/cjnatcop

Climate Policy Initiative. (2018). *Global Climate Finance: An Updated View 2018*. Climate Policy Initiative. Available at: https://climatepolicyinitiative.org/publication/global-climate-finance-an-updated-view-2018/

Cole, W. (2007, June 3). Australia to Launch Carbon Trading Scheme by 2012. *Reuters*. Sydney. Retrieved from http://www.reuters.com/article/2007/06/03/environment-australia-climate-dc-idUSSYD26700820070603

Conelly, J. (2014, July 11). *Obama Will Reject KeystoneXL Pipeline, Sierra Club Boss Predicts*. Retrieved March 19, 2015, from http://blog.seattlepi.com/seattlepolitics/2014/07/11/obama-will-reject-keystonexl-pipeline-sierra-club-boss/#14314101=0

COP17 CMP7. (2011). *Multimedia—Images*. Retrieved March 13, 2015, from http://www.cop17-cmp7durban.com/en/multimedia/image-gallery.html

COP18: Taki Eddine Djeffal, Arab Youth Climate Movement. (2012). Retrieved from http://www.youtube.com/watch?v=q09GTyNAg1E&feature=youtube_gdata_player

Cope, M. (2003). Coding Transcripts and Diaries. In N. Clifford & G. Valentine (Eds.), *Key Methods in Geography* (pp. 445–459). Thousand Oaks: SAGE.

CorpWatch. (2000). *Alternative Summit Opens with Call for Climate Justice*. Retrieved March 14, 2015, from http://www.corpwatch.org/article.php?id=333

Costa, L., Rybski, D., & Kropp, J. P. (2011). A Human Development Framework for CO_2 Reductions. *PLoS One, 6*(12), e29262. https://doi.org/10.1371/journal.pone.0029262.

Cover, R. (1983). Nomos and Narrative. *Harvard Law Review, 97*(4), 4–68.

Cox, K. R. (1998). Spaces of Dependence, Spaces of Engagement and the Politics of Scale, or: Looking for Local Politics. *Political Geography, 17*(1), 1–23. https://doi.org/10.1016/S0962-6298(97)00048-6.

Crang, M., & Cook, I. (2007). *Doing Ethnographies.* London: SAGE.

Cronon, W. (1996). The Trouble with Wilderness: Or, Getting Back to the Wrong Nature. *Environmental History, 1*(1), 7–28.

Crutzen, P. J., & Stoermer, E. F. (2000). Global Change Newsletter. *The Anthropocene, 41*, 17–18.

CSO Equity Review. (2018). *After Paris: Inequality, Fair Shares, and the Climate Emergency.* Manila/London/Cape Town/Washington, et al.: CSO Equity Review Coalition. [civilsocietyreview.org/report2018] https://doi.org/10.6084/m9.figshare.7637669

Dalby, S. (2002). *Environmental Security.* Minneapolis: University of Minnesota Press.

Dalby, S. (2009). *Security and Environmental Change.* Cambridge: Polity.

DARA (Development Assistant Research Associates). (2012). *Climate Vulnerability Monitor, 2nd edition: A Guide to the Cold Calculus of a Hot Planet.* Retrieved from http://daraint.org/climate-vulnerability-monitor/climate-vulnerability-monitor-2012/

Davenport, C., & Landler, M. (2019). Trump Administration Hardens Its Attack on Climate Science. *New York Times* [online], 27 May. Available from: https://www.nytimes.com/2019/05/27/us/politics/trump-climate-science.html

Davis, S. J., & Caldeira, K. (2010). Consumption-Based Accounting of CO_2 Emissions. *Proceedings of the National Academy of Sciences, 107*(12), 5687–5692.

De Schutter, O. (2012, April 24). *Climate Change Is a Human Rights Issue—and That's How We Can Solve It.* Retrieved from http://www.theguardian.com/environment/2012/apr/24/climate-change-human-rights-issue

Delaney, D. (2003). *Law and Nature.* Cambridge: Cambridge University Press.

Delaney, D. (2010). *The Spatial, the Legal and the Pragmatics of World-Making: Nomospheric Investigations.* Abingdon: Routledge.

Delaney, D., Ford, R. T., & Blomley, N. (2001). Preface: Where Is Law. In N. Blomley, D. Delaney, & R. T. Ford (Eds.), *The Legal Geographies Reader: Law, Power and Space* (pp. 13–32). Malden: Blackwell Publishers.

Democracy Now. (2010, December 8). Youth Activists Protest Exclusion from U.N. Climate Summit in Cancún. *Democracy Now.* Retrieved June 10, 2019, from http://www.democracynow.org/2010/12/8/indigenous_youth_groups_hold_climate_justice

Democracy Now. (2017, November 17). Tom Goldtooth: Carbon Trading is "Fraudulent" Scheme to Privatize Air & Forests to Permit Pollution. *Democracy Now.* Available at: https://www.democracynow.org/2017/11/17/tom_goldtooth_carbon_trading_is_fraudulent

Derman, B. B. (2013). Contesting Climate Injustice During COP17. *South African Journal on Human Rights: Climate Change Justice: Narratives, Rights and the Poor, 29*(1), 170–179.

Derman, B. B. (2014). Climate Governance, Justice, and Transnational Civil Society. *Climate Policy, 14*(1), 23–41.

Derman, B. B. (2018). "Climate Change Is About US:" Fence-Line Communities, the NAACP and the Grounding of Climate Justice. In T. Jafry (Ed.), *Routledge Handbook of Climate Justice* (pp. 407–419). London: Routledge.

Derman, B. B. (2019). Legal Geographies. In A. Kobayashi (Ed.), *International Encyclopedia of Human Geography.* Amsterdam: Elsevier.

Dismantle Corporate Power. (2012, October 17). *United Nations: Who Wants to Go Through the Revolving Door?* Retrieved March 4, 2013, from http://www.stopcorporateimpunity. org/?p=2213

Dryzek, J. S., Downes, D., Hunold, C., Schlosberg, D., & Hernes, H.-K. (2003). *Green States and Social Movements: Environmentalism in the United States, United Kingdom, Germany, and Norway: Environmentalism in the United States, United Kingdom, Germany, and Norway.* Oxford: Oxford University Press.

Dulitzky, A. (2006, November 16). Letter to Paul Crowley. *New York Times.* Retrieved from http://graphics8.nytimes.com/packages/pdf/science/16commissionletter.pdf

Earthjustice. (2014). Court Upholds Air Safeguard that Would Prevent Thousands of Deaths. *Earthjustice*, April 15. Available at: https://earthjustice.org/news/press/2014/court-up-holds-air-safeguard-that-would-prevent-thousands-of-deaths

Easterbrook, G. (2007, April). Global Warming: Who Loses—and Who Wins? *The Atlantic.* Retrieved from http://www.theatlantic.com/magazine/archive/2007/04/global-warming-who-loses-and-who-wins/5698/

Ecosocialist Horizons. (2011). *Ecosocialist Prefiguration.* Available at: http://ecosocialisthorizons.com/prefiguration/

Edwards, M. (2009). *Civil Society.* Cambridge/Malden: Polity Press.

Eilperin, J. (2014, July 9). Nebraska Court Date Pushes Final Keystone XL Decision Past the Midterms. *The Washington Post.* Retrieved from http://www.washingtonpost.com/blogs/post-politics/wp/2014/07/09/nebraska-court-date-pushes-final-keystone-xl-decision-past-the-midterms/

EJOLT (Environmental Organizations, Liabilities, and Trade). (2013). *Ecological Debt.* Retrieved from http://www.ejolt.org/2013/05/ecological-debt/

Elden, S. (2009). *Terror and Territory.* Minneapolis: University of Minnesota Press.

Ellison, G. (2014). *New Price Tag for Kalamazoo River Oil Spill Cleanup: Enbridge Says $1.21 Billion.* M Live Michigan. Available at: https://www.mlive.com/news/grand-rapids/index.ssf/2014/11/2010_oil_spill_cost_enbridge_1.html

Environmental Justice Climate Change Initiative. (2002). *10 Principles for Just Climate Change Policies in the United States [EJCC].* Retrieved from http://www.ejcc.org/about/

Environmental Justice Leadership Forum on Climate Change. (2015). *Environmental Justice Leadership Forum on Climate Change.* Retrieved March 17, 2015, from http://www.ejleadershipforum.org/

Escobar, A. (1995). *Encountering Development: The Making and Unmaking of the Third World.* Princeton: Princeton University Press.

Escobar, A. (2008). *Territories of Difference: Place, Movements, Life, Redes.* Durham: Duke University Press.

European Commission. (2015). *Carbon Leakage.* Retrieved March 12, 2015, from http://ec.europa.eu/clima/policies/ets/cap/leakage/index_en.htm

European Union. (2013). *Homepage—Transparency Register.* Retrieved June 26, 2013, from http://ec.europa.eu/transparencyregister/info/homePage.do?locale=en

Ewick, P., & Silbey, S. S. (1991). Conformity, Contestation, and Resistance: An Account of Legal Consciousness. *New England Law Review, 26,* 731.

Extinction Rebellion. (n.d.). *Extinction Rebellion US Demands.* International Extinction Rebellion. Retrieved June 10, 2019, from https://xrebellion.org/xr-us/demands

Extinction Rebellion South Africa. (2019). *Our Demands.* Extinction Rebellion South Africa. Retrieved June 10, 2019, from https://xrebellion.org.za/materialis/our-demands/

Farnworth, E. (2018, December 17). *What Just Happened? 5 Themes from the COP24 Climate Talks in Poland*. World Economic Forum. Retrieved June 10, 2019, from https://www.weforum.org/agenda/2018/12/reflections-on-two-weeks-of-climate-talks/

Featherstone, D. (2008). *Resistance, Space and Political Identities: The Making of Counter-Global Networks*. Hoboken: Wiley-Blackwell.

Featherstone, D. (2011). On Assemblage and Articulation. *Area, 43*(2), 139–142.

Featherstone, D. (2012). *Solidarity: Hidden Histories and Geographies of Internationalism*. London: Zed Books.

Ferguson, A. (1980). *An Essay on the History of Civil Society*. New Brunswick: Transaction Publishers.

Fiorino, D. J. (2006). *The New Environmental Regulation*. Cambridge, MA: The MIT Press.

Fischer, D. (2009, May 29). Climate Change Hits Poor Hardest in U.S. *Scientific American*. Retrieved from http://www.scientificamerican.com/article/climate-change-hits-poor-hardest/

Fisher, D. R. (2010). COP-15 in Copenhagen: How the Merging of Movements Left Civil Society Out in the Cold. *Global Environmental Politics, 10*(2), 11–17.

FOEE (Friends of the Earth Europe). (n.d.). *Climate Justice*. Young Friends of the Earth Europe. Available at: http://www.foeeurope.org/yfoee/climatejustice

FOEI (Friends of the Earth International). (2012). *Reclaim the UN from Corporate Capture*. Retrieved from http://www.foei.org/resources/publications/publications-by-subject/economic-justice-resisting-neoliberalism-publications/reclaim-the-un-from-corporate-capture/

Fontaine, T. (2014, August 1). *Sides Present Health Concerns, Job Loss Worries at EPA Hearings*. Available at: http://triblive.com/news/adminpage/6532254-74/hearings-building-federal

Fontaine, T. (2017, March 27). 'A Perilous Pipeline': Indigenous Groups Line Up Against Keystone XL. *CBC News*. Available at: https://www.cbc.ca/news/indigenous/indigenous-groups-keystonexl-2017-approval-1.4042381

Foucault, M., Burchell, G., Gordon, C., & Miller, P. (1991). *The Foucault Effect: Studies in Governmentality: With Two Lectures by and an Interview with Michel Foucault*. Chicago: University of Chicago Press.

Francis, P. (2015). *Laudato Si: On Care for Our Common Home*. Huntington: Our Sunday Visitor.

Fraser, N. (2011). Marketization, Social Protection, Emancipation: Toward a Neo-Polanyian Conception of Capitalist Crisis. In C. Calhoun & G. Derluiguian (Eds.), *Business as Usual: The Roots of the Global Financial Meltdown* (pp. 137–158). New York: New York University Press.

Fried, L., & Samuelsohn, D. (2009, December 17). Hillary Clinton Pledges $100B for Developing Countries. *The New York Times*. Retrieved from http://www.nytimes.com/cwire/2009/12/17/17climatewire-hillary-clinton-pledges-100b-for-developing-96794.html

Friedman, L. (2012, January 9). *NEGOTIATIONS: Can a New Structure Based on the Notion of "Equity" Replace the Kyoto Pact?* Retrieved February 21, 2015, from http://www.eenews.net/stories/1059958230

Friese, S. (2012). *Qualitative Data Analysis with ATLAS.ti* (1st ed.). London: SAGE.

Friman, M. (2007). *Historical Responsibility in the UNFCCC*. Retrieved from http://www.diva-portal.org/smash/record.jsf?pid=diva2:233089

Galanter, M. (1974). Why the "Haves" Come Out Ahead: Speculations on the Limits of Legal Change. *Law & Society Review, 9*(1), 95–160.

Gardiner, S. M. (2006). A Perfect Moral Storm: Climate Change, Intergenerational Ethics and the Problem of Moral Corruption. *Environmental Values, 15*(3), 397–413.

Gardiner, S. (2011). Climate Justice. In J. S. Dryzek, R. B. Norgaard, & D. Schlosberg (Eds.), *The Oxford Handbook of Climate Change and Society* (pp. 309–322). Oxford: Oxford University Press.

Gardiner, S. M., & Weisbach, D. A. (2016). *Debating Climate Ethics*. Oxford: Oxford University Press.

Gardner, T., & Gordon, J. (2018). U.S. to Conduct Additional Keystone XL Pipeline Review. *Reuters*. Available at: https://www.reuters.com/article/us-usa-keystone-pipeline/u-s-to-conduct-additional-keystone-xl-pipeline-review-idUSKCN1NZ2TH

Gerholdt, R. (2014, December 9). *STATEMENT: Green Climate Fund Reaches $10 Billion Goal: "A Major Milestone"*. Retrieved February 22, 2015, from http://www.wri.org/news/2014/12/statement-green-climate-fund-reaches-10-billion-goal-%E2%80%9C-major-milestone%E2%80%9D

Gersmann, H., & Vidal, J. (2011, November 28). *Q&A: Durban COP17 Climate Talks*. Retrieved March 9, 2013, from http://www.guardian.co.uk/environment/2011/nov/28/durban-cop17-climate-talks

Giddens, A. (2009). *The Politics of Climate Change*. Cambridge: Polity Press.

Gidwani, V. (2009). Civil Society. In D. Gregory, R. Johnston, & G. Pratt (Eds.), *The Dictionary of Human Geography*. Wiley-Blackwell: Hoboken.

Gill, S. (1995). Globalisation, Market Civilisation, and Disciplinary Neoliberalism. *Millenium: Journal of International Studies, 24*(3), 399.

Global Alliance for the Rights of Nature. (2015). *Rights of Nature Advocates at COP17 Durban, South Africa*. Available at: http://therightsofnature.org/rights-of-nature-cop17-durban/

Global Alliance for the Rights of Nature. (n.d.). *The Rights of Nature*. Retrieved February 17, 2015, from http://therightsofnature.org/?page_id=38

Global Humanitarian Forum. (2009). *The Anatomy of a Silent Crisis* (Human Impact Report—Climate Change). Retrieved from http://www.ghf-ge.org/human-impact-report.php

Global Justice Ecology Project. (2015). *About Climate Justice*. Retrieved March 17, 2015, from http://globaljusticeecology.org/climate-justice/

Global Witness. (2011). *Making the Forest Sector Transparent: Annual Transparency Report 2009*. Global Witness. Retrieved March 5, 2013, from http://www.globalwitness.org/library/making-forest-sector-transparent-annual-transparency-report-2009

Gloppen, S., & St. Clair, A. (2012). Climate Change Lawfare. *Social Research: An International Quarterly, 79*(4), 899–930.

Goldenberg, S. (2010a, January 14). *Next Few Weeks Vital for Copenhagen Accord, Says US Climate Change Envoy*. Retrieved November 6, 2014, from http://www.theguardian.com/environment/2010/jan/14/climate-change-us-envoy-copenhagen

Goldenberg, S. (2010b, November 30). *Cancún Climate Change Summit: America Plays Tough*. Retrieved February 20, 2015, from http://www.theguardian.com/environment/2010/nov/30/cancun-climate-change-summit-america

Goldenberg, S. (2014, December 14). *Lima Climate Change Talks End in Agreement—but Who Won?* Retrieved February 24, 2015, from http://www.theguardian.com/environment/2014/dec/14/lima-climate-change-talks-who-won

Goluboff, R. L. (2007). *The Lost Promise of Civil Rights*. Cambridge, MA: Harvard University Press.

Goodale, M., & Merry, S. E. (2007). *The Practice of Human Rights: Tracking Law Between the Global and the Local*. Cambridge: Cambridge University Press.

Goodman, A. (2011, December 14). On Climate Change, the Message Is Simple: Get It Done. *The Guardian*. Available at: http://www.theguardian.com/commentisfree/cifamerica/2011/dec/14/durban-climate-change-conference-2011

Gore, A. (2006). *An Inconvenient Truth: The Planetary Emergency of Global Warming and What We Can Do About It*. Emmaus: Rodale.

Government of Bolivia. (2009). *Commitments for Annex I Parties Under Paragraph 1(b) (i) of the Bali Action Plan: Evaluating Developed Countries Historical Climate Debt to Developing Countries*. Retrieved from http://unfccc.int/resource/docs/2009/awglca6/eng/misc04p01.pdf#page=44

Government of Brazil. (2012). *Cúpula dos Povos—Rio + 20*. Retrieved March 13, 2015, from http://www.rio20.gov.br/clientes/rio20/rio20/sobre_a_rio_mais_20/o-que-e-cupula-dos-povos.html

Government of the Maldives. (2008, September 25). *Submission of the Maldives to the OHCHR Under Human Rights Council Res. 7/23*. Retrieved from http://www.ohchr.org/english/issues/climatechange/docs/submissions/Maldives_Submission.pdf

Government of the United States. (2008, September 25). *Submission of the United States to the OHCHR Under Human Rights Council Res. 7/23*. Retrieved from http://www2.ohchr.org/english/issues/climatechange/docs/submissions/USA.pdf

Graeber, D. (2019). If Politicians Can't Face Climate Change, Extinction Rebellion Will. *New York Times*, [online], 1 May. Available from: https://www.nytimes.com/2019/05/01/opinion/extinction-rebellion-climate-change.html. Accessed 30 May 2019.

Gramsci, A. (1971). *Selections from the Prison Notebooks* (Q. Hoare & G. N. Smith, Eds.). New York: International Publishers Co.

Grassroots Global Justice Alliance. (2011, December 4). *Rural Women's Assembly of Southern Africa Statement to COP17 Leaders*. Retrieved March 17, 2015, from http://ggjalliance.org/node/897

Gray, L. (2010, December 12). *Cancun Climate Change Summit: Bolivians Dance to a Different Beat, but Fail to Derail the Talks*. Retrieved from http://www.telegraph.co.uk/earth/environment/climatechange/8197539/Cancun-climate-change-summit-Bolivians-dance-to-a-different-beat-but-fail-to-derail-the-talks.html

Gray, L. (2011, December 11). *Durban Climate Change: The Agreement Explained*. Retrieved from http://www.telegraph.co.uk/news/earth/environment/climatechange/8949099/Durban-climate-change-the-agreement-explained.html

Great Plains Tar Sands Resistance. (n.d.). *Why Resist Tar Sands?* Retrieved March 19, 2015, from http://gptarsandsresistance.org/why/

Gregory, D. (2004). *The Colonial Present: Afghanistan, Palestine, Iraq*. Oxford: Wiley-Blackwell.

Gregory, D. (2011). Violent Geographies: Law, Violence, and Exception in the Global War Prison. In E. Boehmer & S. Morton (Eds.), *Terror and the Postcolonial: A Concise Companion* (pp. 55–98). Oxford: John Wiley & Sons.

Grossman, Z. (2014, April 23). *The Cowboy-Indian Alliance*. Retrieved March 19, 2015, from http://www.counterpunch.org/2014/04/23/the-cowboy-indian-alliance/

Grossman, Z. (2017). *Unlikely Alliances: Native Nations and White Communities Join to Defend Rural Lands.* Seattle: University of Washington Press.

Grubb, M. (2011). Durban: The Darkest Hour? *Climate Policy, 11*(6), 1269–1271. https://doi.org/10.1080/14693062.2011.628786.

Guardian. (n.d.). The Guardian View on the Lima Climate Change Conference: A Skirmish Before the Real Battle. Retrieved February 24, 2015, from http://www.theguardian.com/commentisfree/2014/dec/14/guardian-view-lima-climate-change-conference-cop-20-skirmish1

Gudynas, E. (2011). Buen Vivir: Today's Tomorrow. *Development, 54*(4), 441–447. https://doi.org/10.1057/dev.2011.86.

Gudynas, E. (2017). Value, Growth, Development: South American Lessons for a New Ecopolitics. *Capitalism Nature Socialism.* https://doi.org/10.1080/10455752.2017.1372502.

Gunningham, N. (2009). Environment Law, Regulation and Governance: Shifting Architectures. *Journal of Environmental Law, 21*(2), 179–212. https://doi.org/10.1093/jel/eqp011.

Haines, A., & Ebi, K. (2019). The Imperative for Climate Action to Protect Health. *New England Journal of Medicine, 380*(3), 263–273. https://doi.org/10.1056/NEJMra1807873.

Hall, S., Massey, D., Rustin, M., & Others. (2013). After Neoliberalism: The Kilburn Manifesto. *Soundings, 1.* Retrieved from http://lwbooks.co.uk/journals/soundings/manifesto.html

Hammel, P. (2018, June 15). In Possible Roadblock for Keystone XL, Pipeline Opponents Gift Land to Ponca. *Omaha World-Herald.* Available at: https://www.omaha.com/news/nebraska/in-possible-roadblock-for-keystone-xl-pipeline-opponents-gift-land/article_cbcf6d2b-1862-5d11-b4dd-5ad81d0378ee.html

Hansen, T. (2011, December 17). "Voices" for Mother Earth Ejected from Climate Convention. *Earth Journalism Network.* Retrieved March 4, 2013, from http://earthjournalism.net/story/%E2%80%98voices%E2%80%99-mother-earth-ejected-climate-convention

Haraway, D. (2015). Anthropocene, Capitalocene, Plantationocene, Chthulucene: Making Kin. *Environmental Humanities, 6*(1), 159–165.

Haraway, D., Ishikawa, N., Gilbert, S. F., Olwig, K., Tsing, A. L., & Bubandt, N. (2016). Anthropologists Are Talking—About the Anthropocene. *Ethnos, 81*(3), 535–564. https://doi.org/10.1080/00141844.2015.1105838.

Hargreaves, S. (2012, May). *COP 17 and Civil Society: The Centre Did Not Hold* (Institute for Global Dialogue, Occassional Paper No. 64). Retrieved from http://www.igd.org.za/publications/occasional-papers

Harrison, D., Foss, A., Klevnas, P., & Radov, D. (2011). Economic Policy Instruments for Reducing Greenhouse Gas Emissions. In J. S. Dryzek, R. B. Norgaard, & D. Schlosberg (Eds.), *The Oxford Handbook of Climate Change and Society* (pp. 521–535). Oxford: Oxford University Press.

Harris-Perry, M. (2017). How to Save the N.A.A.C.P. from Irrelevance. *The New York Times,* May 30. Available at: www.nytimes.com/2017/05/30/opinion/melissa-harris-perry-naacp.html

Hart, G. (2006). Denaturalizing Dispossession: Critical Ethnography in the Age of Resurgent Imperialism. *Antipode, 38*(5), 977–1004.

Hart, G. (2013). Gramsci, Geography, and the Languages of Populism. In E. Michael, H. Gillian, K. Stefan, & L. Alex (Eds.), *Gramsci: Space, Nature, Politics* (pp. 301–320). West Sussex: Wiley-Blackwell.

Harvey, D. (1974). Population, Resources, and the Ideology of Science. *Economic Geography, 50*(3), 256–277.

Harvey, D. (1981). The Spatial Fix–Hegel, von Thunen, and Marx. *Antipode, 13*(3), 1–12.

Harvey, D. (1982). *The Limits to Capital.* Oxford: Blackwell.

Harvey, D. (1996). *Justice, Nature and the Geography of Difference.* Hoboken: Wiley-Blackwell.

Harvey, D. (2005). *The New Imperialism.* Oxford: Oxford University Press.

Harvey, D. (2007). *A Brief History of Neoliberalism.* Oxford: Oxford University Press.

Harvey, D. (2019). *The Significance of China in the World Economy.* (Video recorded lecture) Democracy at Work. Available at: https://www.democracyatwork.info/acc_the_significance_of_china_in_the_global_economy

Harvey, D., & Williams, R. (1995). Militant Particularism and Global Ambition: The Conceptual Politics of Place, Space, and Environment in the Work of Raymond Williams. *Social Text,* (42), 69–98.

Hedges, C. (2015). *Wages of Rebellion: The Moral Imperative of Revolt.* New York: Nation Books.

Hefflinger, M. (2014, April 6). *#NoKXL Crop Art Unveiling.* Retrieved from http://boldnebraska.org/cropart-2/

Heinrich Böll Foundation. (2011). *Mobilising Climate Finance for Africa.* Retrieved March 10, 2013, from http://www.et.boell.org/web/113-266.html

Henson, R. (2014). *The Thinking Person's Guide to Climate Change.* Boston: American Meteorological Society.

Heppenstall, J. (2011, June 2). *Activists Jailed After Copenhagen Climate Summit Protest.* Retrieved October 11, 2013, from http://www.theguardian.com/environment/2011/jun/02/activists-jailed-copenhagen-protest

Herbert, S. K. (1997). *Policing Space: Territoriality and the Los Angeles Police Department.* Minneapolis: University of Minnesota Press.

Herbert, S., Derman, B., & Grobelski, T. (2013). The Regulation of Environmental Space. *Annual Review of Law and Social Science* [online], *9*(1), 227–247. Available from: https://doi.org/10.1146/annurev-lawsocsci-102612-134034. Accessed 30 May 2019.

Hewitt, K. (1983). *Interpretations of Calamity from the Viewpoint of Human Ecology.* Boston: Allen & Unwin.

Heyward, J. C., & Dominic, R. (2016). *Climate Justice in a Non-Ideal World.* Oxford: Oxford University Press.

Hjerpe, M., & Linnér, B.-O. (2010). Functions of COP Side-Events in Climate-Change Governance. *Climate Policy, 10*(2), 167–180. https://doi.org/10.3763/cpol.2008.0617.

How It Works at Occupy Wall Street 9/25/11. (2011). Retrieved from https://www.youtube.com/watch?v=xIK7uxBSAS0&feature=youtube_gdata_player

Howard, B. C. (2014, May 12). West Antarctica Glaciers Collapsing, Adding to Sea-Level Rise. *National Geographic.*

HRC (U.N. Human Rights Council). (2009, March 31). *Human Rights Council Resolution 10/4 Human Rights and Climate Change.* U.N. Symbol A/HRC/RES/10/4. Geneva. Retrieved from http://ap.ohchr.org/documents/E/HRC/resolutions/A_HRC_RES_10_4.pdf

Humphreys, S. (2010). *Human Rights and Climate Change.* Cambridge: Cambridge University Press.

Humphreys, S. (2012). Climate Change and Human Rights: Where Is the Law? In *Address Delivered at the 2012 Rafto Prize Ceremony*. Bergen: Copy on file with the author.

Hussin, I. (2012). Circulations of Law: Colonial Precedents, Contemporary Questions. *Oñati Socio-Legal Series, 2*(7). Retrieved from http://papers.ssrn.com/sol3/papers.cfm?abstract_id=2009231

ICC (Inuit Circumpolar Council). (2005). *Petition to the Inter American Commission on Human Rights Seeking Relief from Violations Resulting from Global Warming Caused by Acts and Omissions of the United States*. ICC Canada. Available from: http://www.inuitcircumpolar.com/inuit-petition-inter-american-commission-on-human-rights-to-oppose-climate-change-caused-by-the-united-states-of-america.html

ICLEI. (n.d.). *Resilient Cities—ICLEI: Home*. Retrieved February 20, 2015, from http://resilient-cities.iclei.org/resilient-cities-hub-site/home/

IEN (Indigenous Environmental Network). (2013). *Campaigns, Programs, Issues*. Indigenous Environmental Network. Available at: http://www.ienearth.org/what-we-do/

IEN (Indigenous Environmental Network). (2018a). *Indigenous Environmental Network*. Available at: www.ienearth.org/

IEN (Indigenous Environmental Network). (2018b, June 12). *Tribes, Landowners, and Climate Groups Expand Campaign to Build Solar Inside Keystone XL Pipeline Route*. Available at http://www.ienearth.org/tribes-landowners-and-climate-groups-expand-campaign-to-build-solar-inside-keystone-xl-pipeline-route/

IEN (Indigenous Environmental Network). (n.d.). *Tar Sands*. Retrieved from http://www.ienearth.org/what-we-do/tar-sands/

International Council on Human Rights Policy. (2009). *Human Rights and Climate Change*. Cambridge University Press. Retrieved from http://eprints.lse.ac.uk/id/eprint/25596

IPCC (Intergovernmental Panel on Climate Change). (2007a). *11.7.2 Carbon Leakage—AR4 WGIII Chapter 11: Mitigation from a Cross-Sectoral Perspective*. Retrieved March 12, 2015, from http://www.ipcc.ch/publications_and_data/ar4/wg3/en/ch11s11-7-2.html

IPCC (Intergovernmental Panel on Climate Change). (2007b). *AR4 SYR Synthesis Report Annexes—Glossary A-D*. Retrieved February 20, 2015, from http://www.ipcc.ch/publications_and_data/ar4/syr/en/annexessglossary-a-d.html

IPCC (Intergovernmental Panel on Climate Change). (2007c). *Fourth Assessment Report: Climate Change 2007*. Retrieved from http://www.ipcc.ch/report/ar4/

IPCC (Intergovernmental Panel on Climate Change). (2012). *Managing the Risks of Extreme Events and Disasters to Advance Climate Change Adaptation*. Retrieved from http://citeseerx.ist.psu.edu/viewdoc/download?doi=10.1.1.258.5480&rep=rep1&type=pdf

IPCC (Intergovernmental Panel on Climate Change). (2013, 2014). *Fifth Assessment Report*. Retrieved from http://www.ipcc.ch/report/ar5/

IPCC (Intergovernmental Panel on Climate Change). (2014). *Climate Change 2014: Impacts, Adaptation, and Vulnerability*. Available from: http://www.ipcc.ch/report/ar5/wg2/

(IPCC) Intergovernmental Panel on Climate Change. (2018). *Global Warming of 1.5°C*. Retrieved from https://www.ipcc.ch/sr15/

Jafry, Tahseen. (Ed.). (2018). *Routledge Handbook of Climate Justice*. Abingdon: Routledge. https://www.amazon.com/Routledge-Handbook-Climate-International-Handbooks/dp/1138689351

James, C. L. R. (2001). *The Black Jacobins: Toussaint L'Ouverture and the San Domingo Revolution*. London: Penguin UK.

Jamieson, A., & Vaughan, A. (2017, March 24). Keystone XL: Trump Issues Permit to Begin Construction of Pipeline. *The Guardian*. Available at: https://www.theguardian.com/environment/2017/mar/24/keystone-xl-pipeline-permit-trump-administration

Jealous, B. (2010). Crisis in Cancún. *The Huffington Post*, December 9. Available at: www.huffingtonpost.com/benjamin-todd-jealous/crisis-in-cancun_b_794688.html

Jealous, B. (2013). This Is the Moment for Action on Climate Change. *Post News Group*, July 17. Available at: www.oaklandpost.org/2013/07/17/oped-this-is-the-moment-for-action-on-climate-change/.

Jeffrey, C. (2013). Geographies of Children and Youth III Alchemists of the Revolution? *Progress in Human Geography, 37*(1), 145–152. https://doi.org/10.1177/0309132511434902.

Johl, A., & Duyck, S. (2012). Promoting Human Rights in the Future Climate Regime. *Ethics, Policy & Environment, 15*(3), 298–302.

Johnson, C. (2012, January 26). *Denmark: Court Upholds Decision in Mass Arrests Case* [web page]. Retrieved March 18, 2015, from http://www.loc.gov/lawweb/servlet/lloc_news?disp3_l205402960_text

Jones, R. (2009). Categories, Borders and Boundaries. *Progress in Human Geography, 33*(2), 174–189.

Jubilee South. (2010). *Towards a Peoples' Agenda on Climate Finance*. Retrieved from http://apmdd.org/component/phocadownload/category/2-ecological-debt-environmental-justice-and-climate-change?download=8:towards-a-peoples-agenda-on-climate-finance-a-part-of-our-platform-for-climate-justice

Kaldor, M., Anheier, H., & Glasius, M. (2003). *Global Civil Society*. Cambridge University Press. Retrieved from http://journals.cambridge.org/production/action/cjoGetFulltext?fu lltextid=1841140

Kaldor, M., Moore, H. L., Selchow, S., & Murray-Leach, T. (2012). *Global Civil Society 2012: Ten Years of Critical Reflection*. Basingstoke: Palgrave Macmillan.

Kanter, J. (2009, December 23). E.U. Blames Others for "Great Failure" on Climate. *The New York Times*. Retrieved from http://www.nytimes.com/2009/12/23/world/europe/23iht-climate.html

Kartha, S., & Erickson, P. (2011). *Comparison of Annex 1 and Non-Annex 1 Pledges Under the Cancun Agreements*. Stockholm Environment Institute. Retrieved from http://www.oxfam.org.nz/resources/onlinereports/SEI-Comparison-of-pledges-Jun2011.pdf

KCRW. (2019, March 7). Is Nuclear Power the Answer to Climate Change? *To the Point*. Retrieved June 10, 2019, from https://www.kcrw.com/news/shows/to-the-point

Keck, M. E., & Sikkink, K. (1998). *Activists Beyond Borders: Advocacy Networks in International Politics*. Ithaca: Cornell University Press.

Keck, M. E., & Sikkink, K. (1999). Transnational Advocacy Networks in International and Regional Politics. *International Social Science Journal, 51*(159), 89–101.

Keim, B., & Macalister, T. (2015, June 15). Shell's Arctic Oil Rig Departs Seattle as "Kayaktivists" Warn of Disaster. *The Guardian*. Retrieved from https://www.theguardian.com/us-news/2015/jun/15/seattle-kayak-activists-detained-blocking-shell-arctic-oil-rig

Khor, M. (2010a). Complex Implications of the Cancun Climate Conference. *Economic and Political Weekly, XLV*(52), 10–15.

Khor, M. (2010b, February 12). *After Copenhagen, the Way Forward*. Retrieved February 21, 2015, from http://www.twn.my/title2/climate/info.service/2010/climate20100212.htm

Khor, M. (2013, November 25). *New Climate Deal on Loss and Damage*. Retrieved February 24, 2015, from http://www.thestar.com.my/Opinion/Columnists/Global-Trends/Profile/Articles/2013/11/25/New-climate-deal-on-loss-and-damage

Kiln. (n.d.). *The Carbon Map*. Retrieved March 15, 2015, from http://www.carbonmap.org/

Klinsky, S., & Brankovic, J. (2018). *The Global Climate Regime and Transitional Justice*. London/New York: Routledge.

Klinsky, S., & Winkler, H. (2014). Equity, Sustainable Development and Climate Policy. *Climate Policy, 14*(1), 1–7. https://doi.org/10.1080/14693062.2014.859352.

Klug, H., & Merry, S. E. (Eds.). (2016). *The New Legal Realism. Vol. 2: Studying Law Globally*. Cambridge: Cambridge University Press.

Knox, J. H. (2009). Linking Human Rights and Climate Change at the United Nations. *Harvard Environmental Law Review, 33*, 477–498.

Kodili, B. C. (2011). *7000 km in 17 days to COP17*. Retrieved March 16, 2015, from http://www.actionaid.se/en/activista/shared/7000km-17-days-cop17

Kolbert, E. (2006). *Field Notes from a Catastrophe: Man, Nature, and Climate Change*. New York: Bloomsbury Pub.

Kolbert, E. (2014). *The Sixth Extinction: An Unnatural History*. New York: Henry Holt and Company.

Koopmans, R. (1999). Political. Opportunity. Structure. Some Splitting to Balance the Lumping. *Sociological Forum, 14*(1), 93–105. https://doi.org/10.2307/685018.

Kovel, J. (2007). *The Enemy of Nature: The End of Capitalism or the End of the World?* London: Zed Books.

Kriesi, H. (1995). The Political Opportunity Structure of New Social Movements: Its Impact on Their Mobilization. In C. J. Jenkins & B. Klandermans (Eds.), *The Politics of Social Protest: Comparative Perspectives on States and Social Movements* (pp. 167–198). Minneapolis: University of Minnesota Press.

Kurtz, H. (2002). The Politics of Environmental Justice as the Politics of Scale: St. James Parish, Louisiana, and the Shintech Siting Controversy. In A. Herod & M. Wright (Eds.), *Geographies of Power: Placing Scale* (pp. 249–273). Oxford: Blackwell.

Kurtz, H. E. (2003). Scale Frames and Counter-Scale Frames: Constructing the Problem of Environmental Injustice. *Political Geography, 22*(8), 887–916. http://doi.org/16/j.polgeo.2003.09.001.

La Via Campesina. (2010a, December 3). *The International Caravan of La Via Campesina Advances for Cancun*. Retrieved March 16, 2015, from http://www.viacampesina.org/en/index.php/actions-and-events-mainmenu-26/-climate-change-and-agrofuels-mainmenu-75/958-the-international-caravan-of-la-via-campesina-advances-for-cancun

La Via Campesina. (2010b, December 10). *Via Campesina Declaration in Cancún: The People Hold Thousands of Solutions in Their Hands*. Retrieved March 16, 2015, from http://www.viacampesina.org/en/index.php/actions-and-events-mainmenu-26/-climate-change-and-agrofuels-mainmenu-75/984-via-campesina-declaration-in-cancun-the-people-hold-thousands-of-solutions-in-their-hands

La Via Campesina. (2011a). *La Vía Campesina Declaration in Cancun*. Retrieved March 16, 2015, from http://viacampesina.org/en/index.php/actions-and-events-mainmenu-26/-climate-change-and-agrofuels-mainmenu-75/992-la-via-campesina-declaration-in-cancun

La Via Campesina. (2011b). *Via Campesina at COP17 in Durban: Industrial Agriculture Heats Up the Planet. Farmers Are Cooling It Down!* Retrieved March 16, 2015, from http://viacampesina.org/en/index.php/actions-and-events-mainmenu-26/-climate-change-and-agrofuels-mainmenu-75/1126-via-campesina-at-cop17-in-durban-industrial-agriculture-heats-up-the-planet-farmers-are-cooling-it-down

Lang, S. (2009). Assessing Advocacy: European Transnational Women's Networks and Gender Mainstreaming. *Social Politics: International Studies in Gender, State & Society, 16*(3), 327.

Latour, B. (1987). *Science in Action: How to Follow Scientists and Engineers Through Society.* Cambridge, MA: Harvard University Press.

Latour, B. (1993). *We Have Never Been Modern.* Cambridge, MA: Harvard University Press.

Latour, B. (2004). *Politics of Nature: How to Bring the Sciences into Democracy.* Cambridge, MA: Harvard University Press.

Latour, B. (2007). *Reassembling the Social: An Introduction to Actor-Network-Theory.* Oxford: OUP.

Lawson, V. (2007). Geographies of Care and Responsibility. *Annals of the Association of American Geographers, 97*(1), 1–11.

Layzer, J. A. (2015). *The Environmental Case: Translating Values into Policy.* Thousand Oaks: Sage.

Lazarus, R. J. (2004). *The Making of Environmental Law.* Chicago: University of Chicago Press.

Lemos, M. C., & Agrawal, A. (2006). Environmental Governance. *Annual Review of Environment and Resources, 31*, 297–325.

Lesnikowski, A., et al. (2016). What Does the Paris Agreement Mean for Adaptation? *Climate Policy* [online], *17*(7), 825–831. Available from https://doi.org/10.1080/14693062.2016.1248889. Accessed 30 May 2019.

Lesnikowski, A., et al. (2017). What Does the Paris Agreement Mean for Adaptation? *Climate Policy, 17*(7), 825–831.

Levin, K., Cashore, B., Bernstein, S., & Auld, G. (2010). Playing It Forward: Path Dependency, Progressive Incrementalism, and the "Super Wicked" Problem of Global Climate Change. In *International Studies Association 48th Annual Convention, February* (Vol. 28).

Limon, M. (2009). Human Rights and Climate Change: Constructing a Case for Political Action. *Harvard Environmental Law Review, 33*, 439–476.

Lister, T. (2010, December 10). *WikiLeaks: Cables Reveal Pessimism in Climate Change Talks.* Retrieved February 20, 2015, from http://www.cnn.com/2010/US/12/10/wikileaks.climate.change/

Lobel, O. (2007). The Paradox of Extra-Legal Activism: Critical Legal Consciousness and Transformative Politics. *Harvard Law Review, 120*, 937–988.

Loftus, A. (2013). Gramsci, Nature, and the Philosophy of Praxis. In E. Michael, H. Gillian, K. Stefan, & L. Alex (Eds.), *Gramsci: Space, Nature, Politics* (pp. 178–196). West Sussex: Wiley-Blackwell.

Lohmann, L. (2008). Carbon Trading, Climate Justice and the Production of Ignorance: Ten Examples. *Development, 51*(3), 359–365. https://doi.org/10.1057/dev.2008.27.

Lynas, M. (2007). *Six Degrees: Our Future on a Hotter Planet.* London: Fourth Estate.

Mace, M. J., & Verheyen, R. (2016). Loss, Damage and Responsibility After COP21: All Options Open for the Paris Agreement. *Review of European, Comparative & International*

Environmental Law [online], *25*(5), 197–214. Available from: https://doi.org/10.1111/reel.12172. Accessed 30 May 2019.

MacNeil, R. (2013). Alternative Climate Policy Pathways in the US. *Climate Policy, 13*(2), 259–276. https://doi.org/10.1080/14693062.2012.714964.

MacNeil, R., & Paterson, M. (2012). Neoliberal Climate Policy: From Market Fetishism to the Developmental State. *Environmental Politics, 21*(2), 230–247.

Mahoney, C. (2008). *Brussels Versus the Beltway: Advocacy in the United States and the European Union.* Washington, DC: Georgetown University Press.

Marks, G., & McAdam, D. (1996). Social Movements and the Changing Structure of Political Opportunity in the European Union 1. *West European Politics, 19*(2), 249–278.

Martin, D. G. (2013). Up Against the Law: Legal Structuring of Political Opportunities in Neighborhood Opposition to Group Home Siting in Massachusetts. *Urban Geography, 34*(4), 523–540. https://doi.org/10.1080/02723638.2013.790640.

Martin, D. G., Scherr, A. W., & City, C. (2009). Making Law, Making Place: Lawyers and the Production of Space. *Progress in Human Geography,* 0309132509337281. https://doi.org/10.1177/0309132509337281

Martuzzi, M., et al. (2004). *The Precautionary Principle: Protecting Public Health, the Environment and the Future of Our Children.* WHO Regional Office for Europe Copenhagen. Available from: http://www.asser.nl/media/2227/cms_eel_96_1_book-precautionary-principle-protecting-public-health-the-environment.pdf

Marx, K. (1976). *The German Ideology: Including Theses on Feuerbach and Introduction to the Critique of Political Economy.* Amherst: Prometheus Books.

Marx, K. (2004). *Capital (Volume 1: A Critique of Political Economy): A Critique of Political Economy.* Lawrence: Digireads.com Publishing.

Marx, K. (2008). *The 18th Brumaire of Louis Bonaparte.* New York: Wildside Press LLC.

Marx, K. (2013). *The Communist Manifesto.* New York: Simon and Schuster.

Mary Robinson Foundation. (n.d.). *Climate Justice.* Retrieved from http://www.mrfcj.org/

Mary Robinson Foundation—Climate Justice. (2011, November 18). *Climate Campaign Travelling Across Africa to COP17.* Retrieved March 16, 2015, from http://www.mrfcj.org/news/climate-campaign-travelling-across-africa-to-cop17.html

Mary Robinson Foundation—Climate Justice. (2017). *Principles of Climate Justice.* Available from: https://www.mrfcj.org/principles-of-climate-justice/

Massey, D. B. (2005a). *For Space.* London: SAGE.

Massey, D. (2005b). London Inside-Out. *Soundings-London-Lawrence and Wishart, 32*, 62.

Massey, D., & Rustin, M. (2013). Rethinking the Neoliberal World Order. In S. Hall, D. Massey, & M. Rustin (Eds.), *After Neoliberalism: The Kilburn Manifesto* (Vol. 1, pp. 191–221). London: Soundings.

May, J., & Daly, E. (2011). *Constitutional Environmental Rights Worldwide* (SSRN Scholarly Paper No. ID 1932779). Rochester: Social Science Research Network.

Mayer, B. (2016). Human Rights in the Paris Agreement. *Climate Law, 6*(1–2), 109–117.

Mayors Climate Protection Center. (2005). *U.S. Conference of Mayors Climate Protection Agreement.* Retrieved February 20, 2015, from http://www.usmayors.org/climateprotection/agreement.htm

McAdam, D. (1982). *Political Process and Black Insurgency, 1930–1970.* Chicago: University of Chicago Press.

McCann, M. W. (1994). *Rights at Work: Pay Equity Reform and the Politics of Legal Mobilization.* Chicago: University of Chicago Press.

McInerney-Lankford, S., Darrow, M., & Rajamani, L. (2011). *Human Rights and Climate Change: A Review of the International Legal Dimensions*. Washington, DC: The World Bank.

McKie, R., & Zee, B. van der. (2009, December 12). *Hundreds Arrested in Copenhagen as Green Protest March Leads to Violence*. Retrieved March 18, 2015, from http://www.theguardian.com/environment/2009/dec/13/hundreds-arrested-in-copenhagen-violence

McLaughlin, P., & Khawaja, M. (2000). The Organizational Dynamics of the US Environmental Movement: Legitimation, Resource Mobilization, and Political Opportunity. *Rural Sociology, 65*(3), 422–439.

Merry, S. E. (2006). *Human Rights and Gender Violence: Translating International Law into Local Justice*. Chicago: University of Chicago Press.

Mertz, E. (2011). Undervaluing Indeterminacy: Translating Social Science into Law. *DePaul Law Review* [online], *60*, 397–412. Available from: http://via.library.depaul.edu/law-review/vol60/iss2/7. Accessed 30 May 2019.

Mertz, E., et al. (Eds.). (2016). *The New Legal Realism: Volume 1: Translating Law-and-Society for Today's Legal Practice*. New York: Cambridge University Press.

Meyer, D. S., & Minkoff, D. C. (2004). Conceptualizing Political Opportunity. *Social Forces, 82*(4), 1457–1492.

Meyer, D. S., & Staggenborg, S. (1996). Movements, Countermovements, and the Structure of Political Opportunity. *American Journal of Sociology, 101*(6), 1628–1660.

Michigan Department of Environmental Quality. (2018). *Oil Spill News and Updates*. https://www.michigan.gov/deq/0,4561,7-135-3313_56784%2D%2D-,00.html

Middleton, N., O'Keefe, P., & Moyo, S. (1993). *The Tears of the Crocodile: From Rio to Reality in the Developing World*. London/Boulder: Pluto Press.

Miller, B. (1994). Political Empowerment, Local–Central State Relations, and Geographically Shifting Political Opportunity Structures: Strategies of the Cambridge, Massachusetts, Peace Movement. *Political Geography, 13*(5), 393–406.

Miller, B. A. (2000). *Geography and Social Movements: Comparing Antinuclear Activism in the Boston Area*. Minneapolis: University of Minnesota Press.

Minnesota Public Radio. (2018). *Rivers of Oil* (audio recording). Available at: https://www.mprnews.org/topic/rivers-of-oil

Minow, M. (1991). *Making All the Difference: Inclusion, Exclusion, and American Law*. Ithaca: Cornell University Press.

Mitchell, T. (2002). *Rule of Experts: Egypt, Techno-Politics, Modernity*. Berkeley: University of California Press.

Mitchell, D. (2003). *The Right to the City: Social Justice and the Fight for Public Space*. New York: Guilford Press.

Mitchell, T. (2011). *Carbon Democracy: Political Power in the Age of Oil*. London: Verso Books.

Mock, B., & Patterson, J. (2014). Want to Support Clean Energy? Fight for Voting Rights. *Grist*, July 25. Available at: http://grist.org/politics/want-to-support-clean-energy-fight-for-voting-rights/

Moe, K. (2013, November 22). "Cowboys and Indians" Camp Together to Build Alliance Against Keystone XL [Article]. Retrieved March 19, 2015, from http://www.yesmagazine.org/peace-justice/cowboys-and-indians-camp-together-to-build-alliance-against-keystone-xl

Moe, K. (2014a, April 23). Photo Essay: "Cowboys and Indians" Against Keystone XL Bring Newfound Unity to DC [Article]. Retrieved March 19, 2015, from http://www.yesmagazine.org/peace-justice/photo-essay-cowboys-and-indians-bring-unity-to-dc

Moe, K. (2014b, April 24). Brought Together by Keystone Pipeline Fight, "Cowboys and Indians" Heal Old Wounds [Article]. Retrieved March 19, 2015, from http://www.yesmagazine.org/peace-justice/brought-together-by-pipeline-fight-cowboys-and-indians-heal-old-wounds

Moe, K. (2014c, May 2). *When Cowboys and Indians Unite—Inside the Unlikely Alliance That Is Remaking the Climate Movement.* Retrieved March 19, 2015, from http://wagingnonviolence.org/feature/cowboys-indians-unite-inside-unlikely-alliance-foretells-victory-climate-movement/

Moore, J. W. (2015). *Capitalism in the Web of Life: Ecology and the Accumulation of Capital.* London: Verso Books.

Mueller, T. (2012). The People's Climate Summit in Cochabamba: A Tragedy in Three Acts. *Ephemera, 12,* 70.

Mufson, S. (2014, April 18). Obama Administration Postpones Decision on Keystone XL Pipeline. *The Washington Post.* Available at: https://www.washingtonpost.com/business/economy/obama-administration-postpones-decision-on-keystone-xl-pipeline/2014/04/18/0c8d9f04-c72a-11e3-8b9a-8e0977a24aeb_story.html?utm_term=.7c8f9793ec7c

Mufson, S., & Eilperin, J. (2017). Trump Seeks to Revive Dakota Access, Keystone XL Oil Pipelines. *The Washington Post.* Available at: https://www.washingtonpost.com/news/energy-environment/wp/2017/01/24/trump-gives-green-light-to-dakota-access-keystone-xl-oil-pipelines/?utm_term=.cff33e1cc15c

NAACP (National Association for the Advancement of Colored People). (2010a). *Climate Justice Initiative Toolkit.* Available at: http://naacp.3cdn.net/112a13293ef36d1c41_o6m6bktqq.pdf

NAACP (National Association for the Advancement of Colored People). (2010b). *Handbook for Advocacy/Programs.* Available at: http://action.naacp.org/page/-/toolkits/Handbook_FINAL.pdf

NAACP (National Association for the Advancement of Colored People). (2011a, July 14). *NAACP Applauds EPA's Cross-State Air Pollution Rule.* Available at: http://www.naacp.org/press/entry/naacp-applauds-epas-cross-state-air-pollution-rule

NAACP (National Association for the Advancement of Colored People). (2011b, August 4). *NAACP Passes Resolution Supporting Strong Clean Air Act.* Available at: https://web.archive.org/web/20140808054436/www.naacp.org/press/entry/naacp-passes-resolution-supporting-strong-clean-air-act

NAACP (National Association for the Advancement of Colored People). (2012). *Coal Blooded Action Toolkit.* Available at: http://action.naacp.org/page/-/Climate/Coal_Blooded_Action_Toolkit_FINAL_FINAL.pdf

NAACP (National Association for the Advancement of Colored People). (2013). *Just Energy Policies: Reducing Pollution and Creating Jobs.* Available at: http://naacp.3cdn.net/8654c676dbfc968f8f_dk7m6j5v0.pdf

NAACP (National Association for the Advancement of Colored People). (2018a). *Climate Justice Initiative.* Available at: www.naacp.org/programs/entry/climate-justice

NAACP (National Association for the Advancement of Colored People). (2018b). *Resource Organizations*. Available at: www.naacp.org/climate-justice-resources/re-source-organizations/

NAACP (National Association for the Advancement of Colored People). (2018c). *The Latest*. Available at: www.naacp.org/latest/?cat=0&topic=47

NAACP (National Association for the Advancement of Colored People). (2018d). *Civil Rights Legislative Report Cards*. Available at: www.naacp.org/report-cards/

NAACP (National Association for the Advancement of Colored People). (n.d.). *NAACP ECJP—YouTube*. Available at: www.youtube.com/user/Katrina2Copenhagen

NAACP (National Association for the Advancement of Colored People), Indigenous Environmental Network, and Little Village Environmental Justice Organization. (2012). *Coal Blooded: Profits Before People*. Available at: www.naacp.org/wp-content/up-loads/2016/04/CoalBlooded.pdf

Naess, A. (1973). The Shallow and the Deep, Long-Range Ecology Movement. A Summary. *Inquiry, 16*(1–4), 95–100. https://doi.org/10.1080/00201747308601682.

Nagar, R., & Raju, S. (2003). Women, NGOs and the Contradictions of Empowerment and Disempowerment: A Conversation. *Antipode, 35*(1), 1–13. https://doi.org/10.1111/1467-8330.00298.

National Resources Defense Council. (2013). *NRDC Policy Basics: Tar Sands*. Available at: https://www.nrdc.org/sites/default/files/policy-basics-tar-sands-FS.pdf

National Transportation Safety Board. (2012). *Enbridge Incorporated Hazardous Liquid Pipeline Rupture and Release Marshall, Michigan July 25, 2010*. Available at: https://www.ntsb.gov/investigations/AccidentReports/Reports/PAR1201.pdf

Ndlovu, M. M. (2013). Qwasha! Climate Justice Community Dialogues Compilation. Vol. 1: Voices from the Streets. *South African Journal on Human Rights: Climate Change Justice: Narratives, Rights and the Poor, 29*(1), 180–191.

Newell, P. (2005). Climate for Change? Civil Society and the Politics of Global Warming. In F. Holland et al. (Eds.), *Global Civil Society Yearbook* (pp. 90–120). London: SAGE.

Newell, P. (2006). *Climate for Change: Non-State Actors and the Global Politics of the Greenhouse*. Cambridge: Cambridge University Press.

Ninth International Conference of American States. (1948). *American Declaration on the Rights and Duties of Man*. Bogota, Colombia (March 30–May 2). Available from: https://www.cidh.oas.org/Basicos/English/Basic2.American%20Declaration.htm

Nixon, R. (2011). *Slow Violence and the Environmentalism of the Poor*. Cambridge, MA: Harvard University Press.

O'Connor, J. (1988). *Capitalism, Nature, Socialism a Theoretical Introduction*. Retrieved from http://www.tandfonline.com/doi/abs/10.1080/10455758809358356

OAS (Organization of American States). (2007). *Human Rights and Global Warming*. IACHR Hearings and Other Public Events. Available from: http://www.oas.org/es/cidh/audiencias/Hearings.aspx?Lang=en&Session=14

OAS (Organization of American States). (2008). *Human Rights and Climate Change in the Americas*. Available from: https://www.oas.org/dil/AGRES_2429.doc

OAS (Organization of American States). (2011). *Annual Report of the Inter-American Commission on Human Rights*. Available from: http://www.oas.org/en/iachr/docs/an-nual/2011/toc.asp

OAS (Organization of American States). (2015). *IACHR Expresses Concern Regarding Effects of Climate Change on Human Rights*. Available from: http://www.oas.org/en/ia-chr/media_center/preleases/2015/140.asp

Occupy Cop17. (2014). *Occupy COP17*. Retrieved March 16, 2015, from https://www.face-book.com/occupyCOP17

OHCHR (Office of the United Nations High Commissioner for Human Rights). (2009). *Report of the Office of the United Nations High Commissioner for Human Rights on the Relationship Between Climate Change and Human Rights*. U.N. Symbol A/HRC/10/61.

OHCHR (Office of the United Nations High Commissioner for Human Rights). (2015). *Understanding Human Rights and Climate Change (Submission to the 21st Conference of Parties to the UNFCCC)*. Geneva, November 27. Available from: http://www.ohchr.org/Documents/Issues/ClimateChange/COP21.pdf

One Million Climate Jobs. (2011). *Climate Jobs Booklet 2011*. Retrieved March 16, 2015, from http://climatejobs.org.za/

Onyango, O. (2011, November 2). *Addis Ababa Meeting Consolidates African Unity on Climate Change*. Retrieved November 18, 2019, from Trans African Caravan of Hope website: http://transafricancaravanofhope.blogspot.com/2011/11/addis-ababa-meeting-consolidates.html

Osaka, S. (2018). How to Disrupt Keystone XL? Solar Panels, Lawsuits, and Ancestral Land. *Grist*. https://grist.org/article/how-to-disrupt-keystone-xl-solar-panels-lawsuits-and-ancestral-land/

Osofsky, H. M. (2006). The Inuit Petition as a Bridge? Beyond Dialectics of Climate Change and Indigenous Peoples' Rights. *American Indian Law Review, 31*(2), 675–697.

Osofsky, H. M. (2008). The Geography of Climate Change Litigation Part II: Narratives of Massachusetts v. EPA. *Chicago Journal of International Law, 8*, 573.

Osofsky, H., & McAllister, L. (2012). *Climate Change Law and Policy*. New York: Aspen Publishers.

Our Power Campaign. (2016). *Our Power Campaign*. Available at: www.ourpowercampaign.org/campaign/

Owe Aku (Bring Back the Way &International Justice Project). (2014, February 6). *Moccasins on the Ground*. Retrieved March 19, 2015, from http://www.oweakuinternational.org/moccasins-on-the-ground.html

Oxfam International. (2015). *Extreme Carbon Inequality*. Oxfam International. Retrieved June 11, 2019, from https://www.oxfam.org/en/research/extreme-carbon-inequality

Parenti, C. (2012). *Tropic of Chaos: Climate Change and the New Geography of Violence*. New York: Nation Books.

Parenti, C. (2014). The 2013 ANTIPODE AAG Lecture The Environment Making State: Territory, Nature, and Value. *Antipode*, n/a–n/a. https://doi.org/10.1111/anti.12134

Paterson, M. (2011). Selling Carbon: From International Climate Regime to Global Carbon Market. In J. S. Dryzek, R. B. Norgaard, & D. Schlosberg (Eds.), *The Oxford Handbook of Climate Change and Society* (pp. 611–624). Oxford: Oxford University Press.

Patterson, J. (2009a, November 17). Natural Disasters, Climate Change Uproot Women of Color (2). *Trouthout*. Available at: http://truth-out.org/archive/component/k2/item/86799:natural-disasters-climate-change-uproot-women-of-color

Patterson, J. (2009b, December 21). The NAACP Offers 10 Lessons From Copenhagen Climate Change Conference. *The Root*. Available at: www.theroot.com/10-lessons-from-copenhagen-1790873906

Patterson, J. (2010a). Your Take: Climate Change Is a Civil Rights Issue. *The Root*. Available at: www.theroot.com/your-take-climate-change-is-a-civil-rights-issue-1790879295

Patterson, J. (2010b). *NAACP Convenes HBCUs in the Gulf Region to Discuss Sustainability Research Agenda.* September 29. Available at: www.naacp.org/latest/hbcu-covening-the-research-agenda-on-the-oil-disaster-and-and-sustaina/

Patterson, J. (2010c), *Vote as If Your Life Depends on It: Especially for US, It Does!* November 1. Available at: https://climatejusticeinitiative.wordpress.com/page/4/

Patterson, J. (2011). *Nature's Fury—Chronicling the Devastating Effects of Climate Change in the US South.* May 8. Available at: https://climatejusticeinitiative.wordpress.com/2011/05/08/nature%E2%80%99s-fury%E2%80%94chronicling-the-devastating-effects-of-climate-change-in-the-us-south/

Patterson, J., & Njamnshi, A. (2011). *From the Bronx to Botswana: Making a Climate Change Connection.* July 30. Available at: https://thegrio.com/2011/07/30/from-balti-more-to-botswana-making-the-climate-change-connection/

Patz, J. A., Campbell-Lendrum, D., Holloway, T., & Foley, J. A. (2005). Impact of Regional Climate Change on Human Health. *Nature, 438*(7066), 310–317.

Patz, J. A., Gibbs, H. K., Foley, J. A., Rogers, J. V., & Smith, K. R. (2007). Climate Change and Global Health: Quantifying a Growing Ethical Crisis. *EcoHealth, 4*(4), 397–405.

Peet, R. (2003). *Unholy Trinity: The IMF, World Bank and WTO.* London/New York: Zed Books.

Peet, R., Robbins, P., & Watts, M. (2010). Global Nature. In R. Peet, P. Robbins, & M. Watts (Eds.), *Global Political Ecology* (pp. 1–48). New York: Routledge.

People's Summit for Environmental and Social Justice. (2012, July 20). *Final Declaration of the People's Summit at Rio+20.* Retrieved from http://www.internationalrivers.org/files/attached-files/rio20_peoplessummit_eng.pdf

Petermann, A. (2011, December 16). *Showdown at the Durban Disaster: Challenging the "Big Green" Patriarchy.* Climate Connections. Retrieved January 4, 2012, from http://climate-connections.org/2011/12/16/showdown-at-the-durban-disaster-challenging-the-big-green-patriarchy/

Peterson, T. C., Alexander, L. V., Allen, M. R., Anel, J. A., Barriopedro, D., Black, M. T., ... others. (2013). Explaining Extreme Events of 2012 from a Climate Perspective. *Bulletin of the American Meteorological Society, 94*(9), S1–S74.

Pettit, J. (2004). Climate Justice: A New Social Movement for Atmospheric Rights. *IDS Bulletin, 35*(3), 102–106. https://doi.org/10.1111/j.1759-5436.2004.tb00142.x.

Pew Charitable Trusts. (2013). *Pew Center on Global Climate Change.* Retrieved July 1, 2013, from http://www.pewtrusts.org/our_work_detail.aspx?id=327744

Phillips, P. (1980). *Marx and Engels on Law and Laws.* Totowa: Barnes & Noble.

Pickering, J., Vanderheiden, S., & Miller, S. (2012). "If Equity's In, We're Out": Scope for Fairness in the Next Global Climate Agreement. *Ethics & International Affairs, 26*(04), 423–443.

Polanyi, K. (2001). *The Great Transformation: The Political and Economic Origins of Our Time* (2nd ed.). Boston: Beacon Press.

Poor People's Campaign. (2018). *Demands.* Available at: www.poorpeoplescampaign.org/demands/

Popke, J. (2008). Geography and Ethics: Non-Representational Encounters, Collective Responsibility and Economic Difference. *Progress in Human Geography, 33*, 81–90.

Popke, J. (2009). Geography and Ethics: Non-Representational Encounters, Collective Responsibility and Economic Difference. *Progress in Human Geography* [online], *33*(1), 81–90. Available from: https://doi.org/10.1177%2F0309132508090441. Accessed 30 May 2019.

Posner, E. A. (2007). Climate Change and International Human Rights Litigation: A Critical Appraisal. *University of Pennsylvania Law Review, 155*(6), 1925–1945.

Prediger, J. (2011, November 2). Green Issues and Greenbacks: Occupy Wall Street Connects the Dots. *Grist.* Retrieved June 10, 2019, from https://grist.org/climate-energy/2011-11-01-green-issues-greenbacks-occupy-wall-street-connects-dots-video/

Primrose, D. (2013). Contesting Capitalism in the Light of the Crisis: A Conversation with David Harvey. *The Journal of Australian Political Economy, 71*, 5.

Protect the Sacred. (n.d.). *About.* Retrieved March 19, 2015, from http://www.protectthesacred.org/about

Putnam, R. D. (2000). *Bowling Alone: The Collapse and Revival of American Community.* New York: Simon & Schuster.

Rabe, B. G. (2010). *Greenhouse Governance: Addressing Climate Change in America.* Washington, DC: Brookings Institution Press.

Rabe, B. G. (2011). *Contested Federalism and American Climate Policy* (SSRN Scholarly Paper No. ID 1902998). Rochester: Social Science Research Network.

Rajamani, L. (2010). The Increasing Currency and Relevance of Rights-Based Perspectives in the International Negotiations on Climate Change. *Journal of Environmental Law, 22*(3), 391–429. https://doi.org/10.1093/jel/eqq020.

Rall, K. (2018, December 10). *Poland's Restrictions on Protest Keep Representatives from Civil Society Groups Barred from Participating in COP24.* Business & Human Rights Resource Centre. Available at: https://www.business-humanrights.org/en/polands-restrictions-on-protest-keep-representatives-from-civil-society-groups-barred-from-participating-in-cop24

Rancière, J., Panagia, D., & Bowlby, R. (2001). Ten Theses on Politics. *Theory & Event, 5*(3). https://doi.org/10.1353/tae.2001.0028.

REDD-Monitor. (2019). *REDD-Monitor.* Retrieved from http://www.redd-monitor.org/

Reinaud, J. (2008). *Climate Policy and Carbon Leakage: Impacts of the European Emissions Trading Scheme on Aluminium.* Paris: International Energy Agency, IEA.

Reject and Protect. (2014a). *Call to Action.* Retrieved March 19, 2015, from http://rejectandprotect.org/call-to-action/

Reject and Protect. (2014b). *Camp Schedule.* Retrieved March 19, 2015, from http://rejectandprotect.org/camp-schedule/

Reject and Protect. (2014c). *Press.* Retrieved March 19, 2015, from http://rejectandprotect.org/press/

Reject and Protect. (2014d). *Reject and Protect: Stop Keystone XL.* Retrieved March 19, 2015, from http://rejectandprotect.org

Renton, A. (2011, January 2). *India's Hidden Climate Change Catastrophe.* Retrieved March 16, 2015, from http://www.independent.co.uk/environment/climate-change/indias-hidden-climate-change-catastrophe-2173995.html

Report of the Office of the United Nations High Commissioner for Human Rights on the Relationship Between Climate Change and Human Rights. (2009). (A/HRC/10/61) [online]. UN Human Rights Council, 15 January. Available from: https://documents-ddsny.un.org/doc/UNDOC/GEN/G09/103/44/PDF/G0910344.pdf?OpenElement. Accessed 30 May 2019.

Revkin, A. C. (2007a, April 1). Poor Nations to Bear Brunt as World Warms. *The New York Times.*

Revkin, A. C. (2007b, April 3). Reports from Four Fronts in the War on Warming. *The New York Times.*

Riahi, K., Rao, S., Krey, V., Cho, C., Chirkov, V., Fischer, G., ... Rafaj, P. (2011). RCP 8.5—A Scenario of Comparatively High Greenhouse Gas Emissions. *Climatic Change, 109*(1), 33.

Rice, J. L. (2010). Climate, Carbon, and Territory: Greenhouse Gas Mitigation in Seattle, Washington. *Annals of the Association of American Geographers, 100*(4), 929–937.

Rising Tide North America. (n.d.). *Rising Tide North America*. Retrieved March 18, 2015, from https://risingtidenorthamerica.org/

Robbins, P. (2004). *Political Ecology: A Critical Introduction*. Malden: Blackwell Pub.

Robbins, P. (2007). *Lawn People: How Grasses, Weeds, and Chemicals Make Us Who We Are*. Philadelphia: Temple University Press.

Robbins, P. (2012). *Political Ecology: A Critical Introduction* (Vol. 20). Chichester: John Wiley & Sons.

Roberts, T. (2012, December 11). *Doha Climate Change Negotiations: Moving Beyond the Dueling Dinosaurs to Bring Together Equity and Ambition*. Retrieved February 21, 2015, from   http://www.brookings.edu/blogs/up-front/posts/2012/12/11-doha-negotiations-roberts

Roberts, J. T., & Parks, B. (2006). *A Climate of Injustice: Global Inequality, North-South Politics, and Climate Policy*. Cambridge: MIT Press.

Roberts, J. T., & Parks, B. C. (2009). Ecologically Unequal Exchange, Ecological Debt, and Climate Justice: The History and Implications of Three Related Ideas for a New Social Movement. *International Journal of Comparative Sociology, 50*(3–4), 385–409.

Robin Hood Tax. (2010). *A Global Financial Transaction Tax for Climate Funding: Investing in Our Collective Future*. Retrieved from http://www.robinhoodtax.org/sites/default/files/RHTC%2520Climate%2520Paper_0.pdf

Robine, J.-M., Cheung, S. L. K., Le Roy, S., Van Oyen, H., Griffiths, C., Michel, J.-P., & Herrmann, F. R. (2008). Death Toll Exceeded 70,000 in Europe During the Summer of 2003. *Comptes Rendus Biologies, 331*(2), 171–178. https://doi.org/10.1016/j.crvi.2007.12.001.

Robinson, M. (2012, February). *Geography of Climate Justice*. Presented at the Annual Meeting of the Association of American Geographers, New York.

Rodríguez-Garavito, C. (2014). The Future of Human Rights: From Gatekeeping to Symbiosis. *SUR-International Journal on Human Rights, 11*(20), 499–510.

Rogelj, J., et al. (2016). Paris Agreement Climate Proposals Need a Boost to Keep Warming Well Below 2 °C. *Nature, 534*(7609), 631–639.

Roland, G. (2014, April 30). *Indigenous Activists Invoke the Sacred as Keystone Pipeline Standoff Continues*. Retrieved March 19, 2015, from http://www.occupy.com/article/indigenous-activists-invoke-sacred-keystone-pipeline-standoff-continues

Rosa Luxemburg Stiftung NYC. (n.d.). *Introducing the Just Transition Research Collaborative. Rosa Luxemburg Stiftung NYC*. Retrieved June 10, 2019, from http://www.rosalux-nyc.org/introducing-the-just-transition-research-collective/

Routledge, P. (2000). 'Our Resistance Will Be As Transnational as Capital': Convergence Space and Strategy in Globalising Resistance. *GeoJournal, 52*(1), 25–33.

Routledge, P. (2003). Convergence Space: Process Geographies of Grassroots Globalization Networks. *Transactions of the Institute of British Geographers, 28*(3), 333–349.

Routledge, P. (2011). Translocal Climate Justice Solidarities. In J. S. Dryzek, R. B. Norgaard, & D. Schlosberg (Eds.), *The Oxford Handbook of Climate Change and Society* (pp. 384–398). Oxford: Oxford University Press.

Routledge, P. (2017). *Space Invaders: Radical Geographies of Protest*. London: Pluto Press.

Routledge, P., & Cumbers, A. (2009). *Global Justice Networks: Geographies of Transnational Solidarity*. Manchester: Manchester University Press.

Routledge, P., Cumbers, A., & Nativel, C. (2007). Grassrooting Network Imaginaries: Relationality, Power, and Mutual Solidarity in Global Justice Networks. *Environment and Planning A, 39*(11), 2575–2592. https://doi.org/10.1068/a38338.

Routledge, P., Cumbers, A., & Derickson, K. D. (2018). States of Just Transition: Realising Climate Justice Through and Against the State. *Geoforum, 88*, 78–86.

Roy, A. (2004). Tide? Or Ivory Snow. *Public Power in the Age of Empire*. http://www.Democracynow.org/2004/8/23/public_power_in_the_age_of

Rubin, H. J., & Rubin, I. (2005). *Qualitative Interviewing: The Art of Hearing Data*. Thousand Oaks: SAGE.

Rural Women's Assembly. (2011). *Memorandum from the Rural Women's Assembly to President Zuma and Minister Nkoane-Mashabane*. Retrieved from https://ruralwomensassembly.wordpress.com/cop17/memorandum/

Rutherford, S. (2007). Green Governmentality: Insights and Opportunities in the Study of Nature's Rule. *Progress in Human Geography, 31*(3), 291–308.

Sachs, W. (2008). Climate Change and Human Rights. *Development, 51*(3), 332–337.

Salamon, L. M., & Sokolowski, S. W. (2004). *Global Civil Society: Dimensions of the Nonprofit Sector*. Baltimore: Johns Hopkins Center for Civil Society Studies.

Salamon, L. M., Sokolowski, S. W., & List, R. (2003). *Global Civil Society: An Overview*. Baltimore: Johns Hopkins Center for Civil Society Studies.

Santos, B. de S., & Rodríguez-Garavito, C. A. (2005). *Law and Globalization from Below: Towards a Cosmopolitan Legality*. Cambridge: Cambridge University Press.

Scheingold, S. A. (1974). *The Politics of Rights*. New Haven: Yale University Press.

Schlanger, Z. (2015, January 20). *Meet the Latina Climate Scientist Michelle Obama Invited to the State of the Union Address*. Retrieved March 19, 2015, from http://www.newsweek.com/meet-latina-climate-scientist-michelle-obama-invited-state-union-address-300877

Schlosberg, D. (2007). *Defining Environmental Justice: Theories, Movements, and Nature*. Retrieved from http://philpapers.org/rec/SCHDEJ

Schroeder, H., & Lovell, H. (2012). The Role of Non-Nation-State Actors and Side Events in the International Climate Negotiations. *Climate Policy, 12*(1), 23–37. https://doi.org/10.1080/14693062.2011.579328.

Schueneman, T. (2010, July 1). *Yvo de Boer Leaves UNFCCC Post "Appalled" by International Inaction*. Retrieved February 20, 2015, from http://globalwarmingisreal.com/2010/07/01/yvo-de-boer-leaves-unfccc-post-appalled-by-international-inaction/

Shannon, M. J. (2010). *NAACP Hosted an Interagency Briefing on July 29 in New Orleans*. August 17. Available at: www.naacp.org/latest/interagency-briefing/

Shearer, C. (2011). *Kivalina: A Climate Change Story*. Chicago: Haymarket Books.

Shift Magazine. (2010). Interview with Erik Swyngedouw: The Post-Politics of Climate Change, (8). Retrieved from http://www.indymedia.org.uk/en/2010/02/446191.html

Shipp, E. R. (2018, April 7). NAACP Poised to Lead Once Again. *The Baltimore Sun.*
    Available at: www.baltimoresun.com/news/opinion/oped/bs-ed-op-0418-shipp-naacp-
    leadership-20180417-story.html
Shiva, V. (2013). *Making Peace with the Earth.* London: Pluto Press.
Silvern, S. E. (1999). Scales of Justice: Law, American Indian Treaty Rights and the Political
    Construction of Scale. *Political Geography, 18*(6), 639–668.
Silverstein, K. (2018). *Questionable Economics Threaten the Keystone XL Pipeline—
    Not Court Rulings. Forbes.* Available at: https://www.forbes.com/sites/kensilver-
    stein/2018/11/11/questionable-economics-threaten-the-keystone-xl-pipeline-not-court-
    rulings/#443b1251134f
Sluyter, M. (2011). *Occupy Wall Street and the Environment: Climate Justice Day.* Human
    Impacts Institute. Retrieved June 10, 2019, from https://www.humanimpactsinstitute.org/
    single-post/2011/11/12/Occupy-Wall-Street-and-the-Environment-Climate-Justice-Day
Smith, N. (2003). *American Empire: Roosevelt's Geographer and the Prelude to
    Globalization.* Berkeley: University of California Press.
Smith, K. (2007). *The Carbon Neutral Myth: Offset Indulgences for Your Climate Sins.*
    Amsterdam: Carbon Trade Watch.
Smith, N. (2008). *Uneven Development: Nature, Capital, and the Production of Space.*
    Athens: University of Georgia Press.
Sparke, M. (2005). *In the Space of Theory: Postfoundational Geographies of the Nation-
    State.* Minneapolis: University of Minnesota Press.
Sparke, M. B. (2006). A Neoliberal Nexus: Economy, Security and the Biopolitics of
    Citizenship on the Border. *Political Geography, 25*(2), 151–180.
Sparke, M. (2012). *Introducing Globalization: Ties, Tensions, and Uneven Integration.*
    Hoboken: John Wiley & Sons.
Spash, C. (2011). Carbon Trading: A Critique. In J. Dryzek, R. Norgaard, & D. Schlosberg
    (Eds.), *The Oxford Handbook of Climate Change and Society* (pp. 550–560). Oxford:
    Oxford University Press.
Spivak, G. C. (1988). *Can the Subaltern Speak?* London: Macmillan.
Stamp Out Poverty. (2009). *Assessing the Alternatives.* Retrieved from http://tilz.tearfund.
    org/~/media/files/tilz/research/sopassessing_web.pdf
Stamp Out Poverty & Institute for Development Studies. (2011. December). *Climate
    Finance.* Retrieved from http://www.gci.org.uk/Documents/Climate_Finance_.pdf
Stand With Africa. (2012). *Stand With Africa: Act Now for Climate Justice.* Retrieved March
    17, 2015, from https://www.facebook.com/pages/Stand-With-Africa/258982300799452
Steffen, W., Crutzen, P. J., & McNeill, J. R. (2007). The Anthropocene: Are Humans Now
    Overwhelming the Great Forces of Nature. *Ambio: A Journal of the Human Environment,
    36*(8), 614–621.
Stephenson, W. (2015). *What We're Fighting for Now Is Each Other: Dispatches from the
    Front Lines of Climate Justice.* Boston: Beacon Press.
Stuart, T. (2019). Sunrise Movement, the Force Behind the Green New Deal, Ramps Up
    Plans for 2020. *Rolling Stone* [online], 1 May. Available from: https://www.rollingstone.
    com/politics/politics-features/sunrise-movement-green-new-deal-2020-828766/.   Ac-
    cessed 30 May 2019.
Submission of the Maldives to the Office of the UN High Commissioner for Human Rights.
    (2008). *Human Rights Council Resolution 7/23. "Human Rights and Climate Change"*

[online]. 25 September 2008. Available from: https://www.ohchr.org/documents/Issues/ClimateChange/Submissions/Maldives_Submission.pdf. Accessed 10 Nov 2017.

Sullivan, P. (2009). *Lift Every Voice: The NAACP and the Making of the Civil Rights Movement*. New York: New Press.

Sunstein, C. R. (2007). Of Montreal and Kyoto: A Tale of Two Protocols. *Harvard Environmental Law Review, 31*(1), 1–65.

Surowiecki, J. (2014, October 6). *Climate Trades*. Retrieved February 24, 2015, from http://www.newyorker.com/magazine/2014/10/13/climate-trades

Sustainable Energy for All. (2013). SE4ALL. Retrieved February 26, 2015, from http://www.se4all.org/

Swyngedouw, E. (1997). Neither Global Nor Local: "Glocalization"and the Politics of Scale. In K. Cox (Ed.), *Spaces of Globalization: Reasserting the Power of the Local, 1* (pp. 137–166). New York/London: Guilford/Longman.

Swyngedouw, E. (1999). Modernity and Hybridity: Nature, Regeneracionismo, and the Production of the Spanish Waterscape, 1890–1930. *Annals of the Association of American Geographers, 89*(3), 443–465. https://doi.org/10.1111/0004-5608.00157.

Swyngedouw, E. (2005). Governance Innovation and the Citizen: The Janus Face of Governance-Beyond-the-State. *Urban Studies, 42*(11), 1991–2006.

Swyngedouw, E. (2007). Impossible "Sustainability" and the Postpolitical Condition. In R. Krueger & D. Gibbs (Eds.), *The Sustainable Development Paradox: Urban Political Economy in the United States and Europe* (pp. 13–40). New York: Guilford Press.

Swyngedouw, E. (2009). The Antinomies of the Postpolitical City: In Search of a Democratic Politics of Environmental Production. *International Journal of Urban and Regional Research, 33*(3), 601–620.

Swyngedouw, E. (2010). Apocalypse Forever? Post-Political Populism and the Spectre of Climate Change. *Theory, Culture & Society, 27*(2–3), 213–232.

Swyngedouw, E. (2013). The Non-Political Politics of Climate Change. *ACME: An International E-Journal for Critical Geographies, 12*(1), 1–8.

Takver. (2011, December 10). *OccupyCOP: Hundreds Protest Inside UN Climate Venue in Durban as Talks Draw to a Close*. Retrieved March 17, 2015, from http://www.indybay.org/newsitems/2011/12/10/18702334.php

Taylor, R. (2000). Watching the Skies: Janus, Auspication, and the Shrine in the Roman Forum. *Memoirs of the American Academy in Rome, 45*, 1–40. https://doi.org/10.2307/4238764.

Taylor, M. (2018, October 26). 'We Have a Duty to Act': Hundreds Ready to Go to Jail over Climate Crisis. *The Guardian*. Available at: https://www.theguardian.com/environment/2018/oct/26/we-have-a-duty-to-act-hundreds-ready-to-go-to-jail-over-climate-crisis

Tejada, C., & Catlin, B. (2013, May 8). *Indigenous Resistance Grows Strong in Keystone XL Battle*. Retrieved March 19, 2015, from http://wagingnonviolence.org/feature/indigenous-resistance-grows-strong-in-keystone-xl-battle/

Terry, G. (2009). No Climate Justice Without Gender Justice: An Overview of the Issues. *Gender and Development, 17*(1), 5–18.

The Danish Institute for Human Rights. (2012, January). *Answers on Best Practices That Promote and Protect the Rights to Freedom of Peaceful Assembly and of Association.*

Retrieved from http://www.ohchr.org/Documents/Issues/FAssociation/Responses2012/NHRI/Denmark.pdf

The Economist. (2011, May 26). Welcome to the Anthropocene. *The Economist*. Retrieved from http://www.economist.com/node/18744401

The Economist. (2012, September 15). Complete Disaster in the Making. *The Economist*. Retrieved from http://www.economist.com/node/21562961

The Economist. (2013, April 20). ETS, RIP? *The Economist*. Retrieved from http://www.economist.com/news/finance-and-economics/21576388-failure-reform-europes-carbon-market-will-reverberate-round-world-ets

The Guardian. (2009, December 16). *In Pictures: Reclaim Power Climate Protest March in Copenhagen*. Retrieved March 18, 2015, from http://www.theguardian.com/environment/gallery/2009/dec/16/reclaim-power-march-copenhagen

The Guardian. (2019). Environment: Extinction Rebellion. *The Guardian*. Available at: https://www.theguardian.com/environment/extinction-rebellion

The Telegraph. (2011, December 9). *Durban Climate Change Conference 2011 Latest*. Retrieved from http://www.telegraph.co.uk/news/earth/environment/climatechange/8916405/Durban-Climate-Change-Conference-2011-latest.html

The White House. (2014, June). *The Health Impacts of Climate Change on Americans*. Retrieved from https://www.whitehouse.gov/sites/default/files/docs/the_health_impacts_of_climate_change_on_americans_final.pdf

Third World Network. (2009). *Climate Debt: A Primer*. Retrieved from http://www.twn.my/announcement/sign-on.letter_climate.dept.htm

Thompson, E. P. (1975). *Whigs and Hunters*. New York: Pantheon Books.

Tincher, S. (2014). WV Shows Dim Efforts in Energy Efficiency. *The (West Virginia) State Journal*, July 24.

Todd, W. A. (n.d.). US Supreme Court Declines to Hear 9th Circuit Decision Rejecting Greenhouse Gas Tort Litigation. *Lexology*. Retrieved February 17, 2015, from http://www.lexology.com/library/detail.aspx?g=21f0d772-3842-42f7-8c82-3d920494960c

Tokar, B. (2010). *Toward Climate Justice*. Porsgrunn: Communalism Press.

Trafford, J. (2019, March 29). Against Green Nationalism. *Open Democracy*. Retrieved June 10, 2019, from https://www.opendemocracy.net/en/opendemocracyuk/against-green-nationalism/

TransCanada Corporation. (2015). *Keystone XL Pipeline Project*. Retrieved March 19, 2015, from http://www.transcanada.com/keystone.html

TransCanada Corporation. (2018). *Keystone XL Pipeline: Route Maps*. Available at http://www.keystone-xl.com/kxl-101/maps/

Transition Network. (2013). *Welcome*. Transition Network. Retrieved February 26, 2015, from https://www.transitionnetwork.org/

Tsing, A. L. (2015). *The Mushroom at the End of the World: On the Possibility of Life in Capitalist Ruins*. Princeton: Princeton University Press.

Turk, A. T. (1976). Law as a Weapon in Social Conflict. *Social Problems, 23*(3), 276–291.

U.S. Global Change Research Program. (2018). *Fourth National Climate Assessment, Volume II: Impacts, Risks, and Adaptation in the United States*. https://nca2018.globalchange.gov/

UNCSD (United Nations Conference on Sustainable Development). (2011). *Green Economy*. Retrieved March 16, 2015, from http://www.uncsd2012.org/greeneconomy.html

UNDP (United Nations Development Program). (2009). *Human Development Report 2009: Overcoming Barriers: Human Mobility and Development.* New York: United Nations Development Program.

UNEP (United Nations Environment Program). (2010). *The Emissions Gap Report 2010.* Retrieved February 21, 2015, from http://www.unep.org/publications/ebooks/emissionsgapreport/

UNEP (United Nations Environment Program). (2014). *Environmental Data Explorer.* Retrieved March 15, 2015, from http://geodata.grid.unep.ch/extras/posters.php

UNEP (United Nations Environment Program). (2017). *The Emissions Gap Report 2017.* Available at: http://www.unep.org/publications/ebooks/emissionsgapreport/

UNEP (United Nations Environment Program). (2018). *Emissions Gap Report 2018.* Retrieved January 2019, from UN Environment website. http://www.unenvironment.org/resources/emissions-gap-report-2018

UNFCCC (United Nations Framework Convention on Climate Change). (1997). *The Kyoto Protocol.* Retrieved from http://unfccc.int/kyoto_protocol/items/2830.php

UNFCCC (United Nations Framework Convention on Climate Change). (2006). *United Nations Framework Convention on Climate Change Handbook.* Bonn: Intergovernmental and Legal Affairs, Climate Change Secretariat. Available from: https://unfccc.int/resource/docs/publications/handbook.pdf

UNFCCC (United Nations Framework Convention on Climate Change). (2007). *Bali Road Map.* Retrieved February 20, 2015, from http://unfccc.int/key_documents/bali_road_map/items/6447.php

UNFCCC (United Nations Framework Convention on Climate Change). (2009). *Copenhagen Accord.* U.N. Symbol FCCC/CP/2009/L.7.

UNFCCC (United Nations Framework Convention on Climate Change). (2011). *Durban Platform for Enhanced Action.* Retrieved February 22, 2015, from http://unfccc.int/bodies/body/6645.php

UNFCCC (United Nations Framework Convention on Climate Change). (2012). *The Doha Climate Gateway.* Retrieved February 24, 2015, from http://unfccc.int/key_steps/doha_climate_gateway/items/7389.php

UNFCCC (United Nations Framework Convention on Climate Change). (2013). Report of the Conference of the Parties on Its Nineteenth Session, Held in Warsaw from 11 to 23 November 2013. Addendum. Part Two: Action Taken by the Conference of the Parties at Its Nineteenth Session. U.N. Symbol FCCC/CP/2013/10/Add.1.

UNFCCC (United Nations Framework Convention on Climate Change). (2014a). *Green Climate Fund.* Retrieved February 22, 2015, from http://unfccc.int/cooperation_and_support/financial_mechanism/green_climate_fund/items/5869.php

UNFCCC (United Nations Framework Convention on Climate Change). (2014b). *Momentum for Change.* Retrieved February 26, 2015, from http://unfccc.int/secretariat/momentum_for_change/items/6214.php

UNFCCC (United Nations Framework Convention on Climate Change). (2014c). *Party Groupings.* Retrieved February 21, 2015, from http://unfccc.int/parties_and_observers/parties/negotiating_groups/items/2714.php

UNFCCC (United Nations Framework Convention on Climate Change). (2019). *REDD+ Home.* Retrieved June 26, 2019, from https://redd.unfccc.int/

United Nations. (1992). *United Nations Framework Convention on Climate Change.* U.N. Symbol A/AC.237/18.

United Nations. (2011). *International Year for People of African Descent 2011.* Retrieved March 17, 2015, from http://www.un.org/en/events/iypad2011/global.shtml

US Department of State. (2013, October 22). *The Shape of a New International Climate Agreement.* Retrieved February 21, 2015, from http://www.state.gov/e/oes/rls/remarks/2013/215720.htm

Van Der Heijden, H.-A. (2006). Globalization, Environmental Movements, and International Political Opportunity Structures. *Organization & Environment, 19*(1), 28–45.

Vidal, J. (2009, December 18). *Rich and Poor Countries Blame Each Other for Failure of Copenhagen Deal.* Retrieved March 18, 2015, from http://www.theguardian.com/environment/2009/dec/19/copenhagen-blame-game

Vidal, J. (2010, February 10). *Yvo de Boer Steps Down as UN Climate Chief to Work for Accountants KPMG.* Retrieved February 20, 2015, from http://www.theguardian.com/environment/2010/feb/18/yvo-de-boer-climate-change

Vidal, J. (2011, December 5). *Durban Climate Talks: Day Eight Diary.* Retrieved March 13, 2015, from http://www.theguardian.com/environment/blog/2011/dec/05/durban-climate-talks-day-eight-diary

Vidal, J. (2012, December 4). *Doha Climate Conference Diary: Youth Activists Bring Energy and Urgency.* Retrieved March 5, 2013, from http://www.guardian.co.uk/environment/blog/2012/dec/04/doha-climate-conference-diary

Vidal, J., Stratton, A., & Copenhagen, S. G. in. (2009). *Low Targets, Goals Dropped: Copenhagen Ends in Failure.* Retrieved February 10, 2015, from http://www.theguardian.com/environment/2009/dec/18/copenhagen-deal

Vincent, A. (1993). Marx and Law. *Journal of Law and Society, 20*(4), 371–397.

Wainwright, J., & Mann, G. (2013). Climate Leviathan. *Antipode, 45*(1), 1–22. https://doi.org/10.1111/j.1467-8330.2012.01018.x.

Wainwright, J., & Mann, G. (2018). *Climate Leviathan: A Political Theory of Our Planetary Future.* London/Brooklyn: Verso.

Wallace, R. (2014, January). *Whipsaw of Damocles: Are Climate Change and Pandemic Influenza Related?* Presented at the Simpson Center for the Humanities, University of Washington.

Wallace-Wells, D. (2019). *The Uninhabitable Earth: Life After Warming.* New York: Crown/Archetype.

Walsh, B. (2010, December 10). Climate: Why the U.S. Is Bargaining So Hard at Cancún. *Time.*

Watt-Cloutier, S. (2018). *The Right to Be Cold: One Woman's Fight to Protect the Arctic and Save the Planet from Climate Change.* Minneapolis: University of Minnesota Press.

WBGU (German Advisory Council on Global Change). (2009). *Solving the Climate Dilemma: The Budget Approach.* Berlin: German Advisory Council on Global Change.

Weart, S. R. (2008). *The Discovery of Global Warming: Revised and Expanded Edition* (Rev. ed.). Cambridge, MA: Harvard University Press.

Weber, M., Rheinstein, M., & Shils, E. A. (1954). *Max Weber on Law in Economy and Society.* Cambridge, MA: Harvard University Press.

Werksman, J. (2010, December 17). *Q&A: The Legal Character and Legitimacy of the Cancun Agreements.* Retrieved February 21, 2015, from http://www.wri.org/blog/2010/12/qa-legal-character-and-legitimacy-cancun-agreements

Werksman, J. (2011). *The Challenge of Legal Form at the Durban Climate Talks.* World Resources Institute. Available from: http://www.wri.org/blog/2011/11/challenge-legal-form-durban-climate-talks

Wernick, A. (2018, 12 May). The "Valve Turners": Activists Faced Jail Time to Briefly Stop the Flow of Canadian Crude Oil. *Public Radio International*. Retrieved June 10, 2019, from https://www.pri.org/stories/2018-05-12/valve-turners-activists-faced-jail-time-briefly-stop-flow-canadian-crude-oil

WHO (World Health Organization). (2004). *Comparative Quantification of Health Risks: Global and Regional Burden of Disease Attributable to Selected Major Risk Factors*. Geneva: World Health Organization. Retrieved from http://www.who.int/healthinfo/global_burden_disease/cra/en/

WHO (World Health Organization). (2005). *Deaths from Climate Change* (map). Available at: http://www.who.int/heli/risks/climate/climatechange/en/

WHO (World Health Organization). (2007). *Climate Change: Quantifying the Health Impact at National and Local Levels*. Geneva: World Health Organization. Available at: https://apps.who.int/iris/bitstream/handle/10665/43708/9789241595674_eng.pdf?sequence=1

WHO (World Health Organization). (2014). *Quantitative Risk Assessment of the Effects of Climate Change on Selected Causes of Death, 2030s and 2050s*. Geneva: World Health Organization.

WHO (World Health Organization). (2019). *Climate Change*. WHO. Available from: http://www.who.int/heli/risks/climate/climatechange/en/

WHO Task Group. (1990). *Potential Health Effects of Climatic Change*. WHO Symbol WHO/PEP/90/10. Geneva: World Health Organization.

Wildcat, D. R. (2013). Introduction: Climate Change and Indigenous Peoples of the USA. *Climatic Change, 120*(3), 509–515. https://doi.org/10.1007/s10584-013-0849-6.

Wilson Center. (2013). *Energy, Climate Change, and Security: Connecting the Dots*. Wilson Center. Retrieved July 1, 2013, from http://www.wilsoncenter.org/dialogue-program/energy-climate-change-and-security-connecting-the-dots

Winkler, H., Jayaraman, T., Pan, J., de Oliveira, A. S., Zhang, Y., Sant, G., … Raubenheimer, S. (2011). Equitable Access to Sustainable Development. In *Contribution to the Body of Scientific Knowledge: A Paper by Experts from BASIC Countries, BASIC Expert Group: Beijing, Brasilia, Cape Town and Mumbai*. Retrieved from http://www.mapsprogramme.org/wp-content/uploads/Basic_EASD_Experts_Paper.pdf

Woodward, K., Jones, J. P., III, & Marston, S. A. (2010). Of Eagles and Flies: Orientations Toward the Site. *Area, 42*(3), 271–280.

Workman, J. (2019, January 25). "Our House Is on Fire." 16 Year-Old Greta Thunberg Wants Action. *World Economic Forum*. Retrieved June 10, 2019, from https://www.weforum.org/agenda/2019/01/our-house-is-on-fire-16-year-old-greta-thunberg-speaks-truth-to-power/

World Development Movement & Jubilee Debt Campaign. (2009, November). *The Climate Debt Crisis: Why Paying Our Dues Is Essential for Tackling Climate Change*. Retrieved from http://slettgjelda.no/assets/docs/Rapporter/climatedebtcrisis1.pdf

World Economic Forum. (2013). *The Ways and Means to Unlock Private Finance for Green Growth*. Available at: http://www3.weforum.org/docs/WEF_GreenInvestment_Report_2013.pdf

WPCCC. (2010a). *People's Agreement*. Retrieved March 4, 2013, from https://pwccc.wordpress.com/support/

WPCCC. (2010b). *Rights of Mother Earth*. Retrieved March 4, 2013, from http://pwccc.wordpress.com/programa/

WRI (World Resources Institute). (2013). *New Report Connects 2012 Extreme Weather Events to Human-Caused Climate Change*. Retrieved February 20, 2015, from http://www.wri.org/blog/2013/09/new-report-connects-2012-extreme-weather-events-human-caused-climate-change

Zajak, S. (2014). *Pathways of Transnational Activism: A Conceptual Framework* (MPIfG Discussion Paper). Retrieved from http://www.econstor.eu/handle/10419/96143

Zee, B. van der. (2010, December 16). *Danish Police Ordered to Compensate Climate Protesters*. Retrieved March 18, 2015, from http://www.theguardian.com/environment/2010/dec/16/danish-police-protesters-compensation

Zemans, F. K. (1983). Legal Mobilization: The Neglected Role of the Law in the Political System. *The American Political Science Review, 77*(3), 690–703.

Legal Cases

Massachusetts v. EPA, 127 S. Ct. 1438 (Supreme Court 2007).

Native Village of Kivalina v. ExxonMobil Corp. F. Supp. 2d 663, 863 (Dist. Ct., ND California, Oakland Div. 2009).

Petition Requesting for Investigation of the Responsibility of the Carbon Majors for Human Rights Violations or Threats of Violations Resulting from the Impacts of Climate Change. CHR-NI-2016-0001 (Republic of the Philippines Commission on Human Rights, Quezon City. 2016).

Urgenda Foundation v. The State of the Netherlands (Ministry of Infrastructure and the Environment), C/09/456689 / HA ZA 13–1396 (Dist. Ct., The Hague. 2015).

Index

© The Author(s) 2020
B. B. Derman, *Struggles for Climate Justice*,
https://doi.org/10.1007/978-3-030-27965-3

Printed in the United States
By Bookmasters